The Geometry of Walker Manifolds

Synthesis Lectures on Mathematics and Statistics

Editor
Steve G. Krantz, *Washington University, St. Louis*

The Geometry of Walker Manifolds

M. Brozos-Vázquez, E. García-Río, P. Gilkey, S. Nikčević, R. Vázquez-Lorenzo

www.morganclaypool.com

ISBN: 9781598298192   paperback
ISBN: 9781598298208   ebook

DOI 10.2200/S00197ED1V01Y200906MAS005

A Publication in the Morgan & Claypool Publishers series
SYNTHESIS LECTURES ON MATHEMATICS AND STATISTICS

Lecture #5
Series Editor: Steve G. Krantz, Washington University, St. Louis

Series ISSN
Synthesis Lectures on Mathematics and Statistics
Print 1930-1743   Electronic 1930-1751

The Geometry of
Walker Manifolds

Miguel Brozos-Vázquez
Department of Mathematics, University of Coruña, Spain

Eduardo García-Río
Department of Geometry and Topology
Faculty of Mathematics, University of Santiago de Compostela, Spain

Peter Gilkey
Mathematics Department, University of Oregon, Eugene, OR, USA

Stana Nikčević
Mathematical Institute, Belgrade, Serbia

Ramón Vázques-Lorenzo
Faculty of Mathematics, University of Santiago de Compostela, Spain

SYNTHESIS LECTURES ON MATHEMATICS AND STATISTICS #5

MORGAN & CLAYPOOL PUBLISHERS

ABSTRACT

This book, which focuses on the study of curvature, is an introduction to various aspects of pseudo-Riemannian geometry. We shall use Walker manifolds (pseudo-Riemannian manifolds which admit a non-trivial parallel null plane field) to exemplify some of the main differences between the geometry of Riemannian manifolds and the geometry of pseudo-Riemannian manifolds and thereby illustrate phenomena in pseudo-Riemannian geometry that are quite different from those which occur in Riemannian geometry, i.e. for indefinite as opposed to positive definite metrics.

Indefinite metrics are important in many diverse physical contexts: classical cosmological models (general relativity) and string theory to name but two. Walker manifolds appear naturally in numerous physical settings and provide examples of extremal mathematical situations as will be discussed presently.

To describe the geometry of a pseudo-Riemannian manifold, one must first understand the curvature of the manifold. We shall analyze a wide variety of curvature properties and we shall derive both geometrical and topological results. Special attention will be paid to manifolds of dimension 3 as these are quite tractable. We then pass to the 4 dimensional setting as a gateway to higher dimensions.

Since the book is aimed at a very general audience (and in particular to an advanced undergraduate or to a beginning graduate student), no more than a basic course in differential geometry is required in the way of background. To keep our treatment as self-contained as possible, we shall begin with two elementary chapters that provide an introduction to basic aspects of pseudo-Riemannian geometry before beginning on our study of Walker geometry. An extensive bibliography is provided for further reading.

Math subject classifications
Primary: 53B20 – (PACS: 02.40.Hw)
Secondary: 32Q15, 51F25, 51P05, 53B30, 53C50, 53C80, 58A30, 83F05, 85A04

KEYWORDS

affine connection, affine surface, almost Hermitian, almost Kaehler, Christoffel symbols, Codazzi Ricci tensor, commuting curvature model, conformally flat, conformally Kaehler, conformally Osserman, contact Walker manifold, curvature commuting, cyclic parallel Ricci tensor, Einstein, flat connection, foliated Walker manifold, Gray identity, geometry of the curvature operator, homogeneous space, hyper Hermitian, hyper-Kaehler, Ivanov-Petrova, Jacobi operator, Levi-Civita connection, locally symmetric, Lorentzian, Nijenhuis tensor, nilpotent Walker manifold, null distribution, Osserman curvature model, para-Hermitian, para-Kaehler, parallel null distribution, projectively flat, Ricci anti-symmetric, Ricci curvature, Ricci flat, scalar curvature, Riemannian extension, torsion free connection, Schouten tensor, sectional curvature, skew-symetric curvature operator, Tricerri-Vanhecke decomposition, Vaisman manifold, vanishing scalar invariants, Walker coordinates, Walker manifold, Weyl curvature, Weyl scalar invariants

Contents

Preface

Much of the early research in differential geometry was expressed in terms of local coordinates and often dealt with purely local phenomena. However, in the years following the second world war, the relationships between the geometry of a manifold and its underlying topological structure began to be explored.

Among the many notable results of this period is a theorem of de Rham [229] in 1952 which concerns the structure of a complete connected Riemannian manifold M of dimension m on which there is defined a field $\mathcal{D}$ of tangent k-planes which is *parallel* with respect to the Levi-Civita connection of M. Here, of course, one assumes $0 < k < m$ so the question is non-trivial. The complementary tangent field $\mathcal{D}^\perp$ of $(m - k)$-planes is parallel as well. Since these two distributions are necessarily integrable, they define complementary foliations $\mathcal{F}$ and $\mathcal{F}^\perp$. Such a structure has topological implications. De Rham showed that if such a manifold M is simply connected, then M is isometric to the orthogonal product $F \times F^\perp$ of any two leaves

$$F \in \mathcal{F} \qquad \text{and} \qquad F^\perp \in \mathcal{F}^\perp.$$

At this time, Walker [256, 257] studied the same problem from the point of view of fiber bundles, motivated by an earlier attempt by Thomas [250] to obtain a global "product theorem" [135]. Walker considered the case where the leaves of the foliation $\mathcal{F}$ fiber M, giving a definitive description of the structure of such bundles. The two pieces of work are closely related, although de Rham's formulation tackles the general case explicitly.

Walker was interested in this problem in the pseudo-Riemannian context where the inner product in question has signature (p, q) for p and q both non-zero. If $\mathcal{D} \cap \mathcal{D}^\perp$ is non-trivial, or, equivalently, if the induced metric on the distribution $\mathcal{D}$ is degenerate, then the situation is quite different. Walker recognized that a full understanding of the global structure of M in the general case was out of reach, and instead concentrated on devising coordinate systems in which the metric tensor took a simple *canonical form*, yielding information on the pseudo-group of coordinate transformations for the structure, and hence some insight into global questions. In [254] he began to explore the new features of the pseudo-Riemannian case. In [255] he studied parallel fields of null k-planes, and obtained a local canonical form for the metric in this setting. Henceforth, we will say that $\mathcal{M} = (M, g)$ is a *Walker manifold* and that g is a *Walker metric* if there is a non-trivial totally isotropic parallel distribution; the manifold or metric is said to be *strict* if the distribution in question is generated by parallel vector fields.

The case in which $m = 2k$ is of special interest, and is the subject of [221]. In 1964 Wu [261] considered the pseudo-Riemannian case, and succeeded in showing that de Rham's theorem still holds, with the obvious modifications to the statement. In two later papers [258, 259] there are global results on the existence of affine connections with respect to which one or more given distributions are parallel. We refer to [121] and the references contained therein for additional examples which use Walker's results to obtain further information of a global character.

There is much recent investigation into Walker geometry in the mathematical physics literature. Lorentzian Walker manifolds have been studied extensively in the physics literature since they constitute the background metric of the pp-wave models [2, 179, 181, 200]; a pp-wave spacetime admits a covariantly constant null vector field U and therefore it is trivially recurrent, i.e. one has the relation:

$$\nabla U = \omega \otimes U \quad \text{for some 1 form} \quad \omega.$$

Lorentzian Walker manifolds present many specific features both from the physical and geometric viewpoints [67, 80, 190, 225]. We also refer to related work of Hall [165] and of Hall and da Costa [166] for generalized Lorentzian Walker manifolds – these are spacetimes admitting a non-zero vector field n^ℓ satisfying

$$R_{ijk\ell}n^\ell = 0$$

or admitting a rank 2-symmetric or anti-symmetric tensor H_{ab} with $\nabla H = 0$. We also refer to the discussion in [93, 94, 95, 103, 197, 198, 199] for other work in the mathematical physics literature.

Pseudo-Riemannian metrics of signature other than Lorentzian have received considerable attention in mathematical physics since the work of Ooguri and Vafa [216] on $\mathcal{N} = 2$ strings [22, 82, 176, 195]. Sahni and Shtanov [230] have provided applications of pseudo-Riemannian metrics in braneworld cosmology.

Walker manifolds constitute the underlying structure of many strictly pseudo-Riemannian situations with no Riemannian counterpart: indecomposable (but not irreducible) holonomy [23], Einstein hypersurfaces with nilpotent shape operators [194] or some classes of Osserman metrics which are not symmetric [106] are typical examples. Walker manifolds have also been considered in general relativity in the study of $\mathfrak{hh}$ spaces [47, 119]. Moreover, the fact that para-Kaehler and hyper-symplectic metrics are necessarily of Walker type motivates the consideration of such metrics in connection with almost para-Hermitian structures.

In this book, we study the geometry of Walker manifolds. We will often omit technical proofs based on lengthy computations in the interests of brevity and refer the reader instead to the original references in the subject. Many of these calculations are long and straightforward; computer algebra programs are useful in these settings.

Here is a brief outline to the book. As our book is intended to be accessible to a wide audience – and in particular to an advanced undergraduate or to a beginning graduate student – we begin with two Chapters which are elementary and present an introduction to the subject. Chapter 1 deals with algebraic preliminaries. It is often useful to work first in the algebraic context and then pass subsequently to the geometrical setting. We introduce curvature models, discuss the Jacobi and the skew symmetric curvature operators, and examine questions related to the spectral geometry or commutativity properties of these operators. Chapter 2 deals with the geometric context. We introduce basic manifold theory, pseudo-Riemannian geometry, pseudo-Hermitian and para-Hermitian structures.

In Chapter 3 we introduce Walker geometry. We treat Walker coordinates, introduce several different families of Walker manifolds, and treat the theory of Riemannian extensions. In Chapter 4, we specialize to the 3 dimensional setting discussing adapted coordinates and local forms for the Ricci and curvature tensors of a Walker manifold. We explore some features of Walker manifolds with non-zero scalar curvature, examine strict Walker manifolds, and discuss curvature homogeneous Lorentzian manifolds. Chapter 5 treats 4 dimensional Walker geometry. Formulas are given for the Levi-Civita connection, the curvature tensor, the Ricci tensor, and the Einstein equations in the context of Walker geometry which are central to our subsequent discussion.

Chapter 6 treats the spectral geometry of the curvature tensor. One imposes certain natural algebraic conditions on the curvature tensor and examines the associated geometric consequences. We examine Osserman geometry in both the context of Walker geometry and in the more general context of manifolds of signature $(2, 2)$. We also treat Ivanov–Petrova geometry and analyze Riemannian extensions of affine surfaces with several properties on the curvature operators. Chapter 7 treats Hermitian structures in dimension 4. We give a local description of proper almost Hermitian Walker structures that are locally conformally Kaehler, self-dual, ⋆-Einstein or Einstein. We also show that any proper almost Hermitian structure on a Walker manifold of dimension 4 is isotropic Kaehler. Moreover, a local description of proper almost Kaehler Walker structures that are self-dual, ⋆-Einstein or Einstein is given and it is proved that any proper strictly almost Kaehler Einstein structure is self-dual, Ricci flat and ⋆-Ricci flat. This is used to supply examples of flat indefinite non-Kaehler almost Kaehler structures.

Chapter 8 deals with special Walker manifolds. Walker manifolds of dimension 4 are parametrized by a triple of functions (a, b, c). We shall set two of the parameters to zero so the Walker manifold is defined by a single function. We examine the curvature tensor of such manifolds, study proper complex structures, determine the eigenvalues of the Weyl operator, and examine commutativity properties of the curvature. We examine the conformal Osserman operator, study geodesic completeness, Ricci blowup, and curvature homogeneity.

The book concludes with a lengthy bibliography. Whilst a complete bibliography is impossible, we have attempted to list many of the major works in the field. As this book is intended for a wide audience, we have also attempted to outline some (but clearly not all) of the history of some of the developments in the field in each Chapter. A glossary with the main notational conventions employed is provided at the end of the book.

Acknowledgments: The research of all of the authors was partially supported by Project MTM2006-01432 (Spain); the research of S. Nikčević was partially supported by Project 144032 (Srbija).

M. Brozos-Vázquez, E. García-Río, P. Gilkey, S. Nikčević, R. Vázquez-Lorenzo
May 2009

CHAPTER 1

Basic Algebraic Notions

1.1 INTRODUCTION

The first two Chapters are intended to be elementary and serve as an introduction to the subject for the reader who is perhaps not familiar with the matters under consideration. It is often convenient to work in a purely algebraic setting and then pass to a more geometrical one; questions of geometric realizability then arise as there are often algebraic structures which have no corresponding geometrical analogues. In this Chapter, we introduce the algebraic structures that we will be using; the reader may want to read the next Chapter (which deals with the corresponding geometric structures) before reading this Chapter, as these structures provide the motivation behind those considered here. It is, however, convenient from a purely practical matter to discuss the algebraic context first.

Here is a brief outline to this Chapter. In Section 1.2, we present a brief historical outline. In Section 1.3, we present some basic algebraic notions. In Section 1.3.1, the Jordan normal form is introduced; it will play an important role all along the book and especially in Section 1.4.2. In Section 1.3.2, we introduce indefinite signature inner products; in Section 1.3.3, we discuss algebraic curvature tensors. We then pass to the complex context introducing Hermitian and para-Hermitian structures in Section 1.3.4. In Section 1.3.5, we discuss the Jacobi and skew symmetric curvature operators, and in Section 1.3.6, sectional curvature and Ricci curvature are treated. The latter leads to a curvature decomposition of the space of algebraic curvature tensors that we present in Section 1.3.7; special attention is paid to the Tricerri-Vanhecke irreducible decomposition in the Hermitian context. The Weyl tensor W is presented as well; in 4 dimensions there is an additional splitting

$$W = W^+ \oplus W^-$$

which gives rise to the notions of self-dual and anti-self-dual as is discussed in Section 1.3.8.

Section 1.4 deals with properties of the following natural operators associated to the curvature tensor:

$$\mathcal{J} \quad \text{the Jacobi operator,}$$
$$\mathcal{R} \quad \text{the curvature operator,}$$
$$\rho \quad \text{the Ricci operator.}$$

In Section 1.4.1, the notion of Osserman tensor is defined; in Section 1.4.2, this notion is specialized to signature $(2, 2)$ and a basic algebraic classification result is given. The conformally Osserman condition is treated as well in this section. In Section 1.4.3, Ivanov–Petrova models are treated. In Section 1.4.4, Osserman Ivanov–Petrova models of signature $(2, 2)$ are discussed. In Section 1.4.5, curvature commutativity properties are presented.

1.2   A HISTORICAL PERSPECTIVE IN THE ALGEBRAIC CONTEXT

Geometric information about a pseudo-Riemannian manifold (M, g) is essentially encoded by the curvature tensor $R \in \otimes^4 T^* M$. Hence, a central problem in differential geometry is to relate algebraic properties of the curvature tensor to the underlying geometry of the manifold. Because the full curvature tensor is a difficult object to study, the investigation usually focuses on different objects associated to the curvature tensor. Algebraic properties of curvature operators have been extensively investigated, with special attention being paid to both the spectrum of such operators (the Jacobi operator and the skew symmetric curvature operator being typical examples [172, 218]), and to the existence of commutativity relations between the Ricci operator and different curvature operators [27, 169, 237]. Boundedness properties have also been examined [29].

Commutativity properties of curvature operators have been systematically investigated during the last years. The skew symmetric curvature operator and the Jacobi operator were first studied in the Riemannian setting for hypersurfaces in $\mathbb{R}^{m+1}$ [253] and then subsequently studied in the general pseudo-Riemannian context in [53, 54, 55]. Other commutativity relations between the Ricci, the Jacobi and the skew symmetric curvature operators have also been considered in the literature. We refer to [48, 50, 150] for more information.

The skew symmetric curvature operator can be regarded as the part of the curvature tensor describing the behavior of circles [171]. Geodesics and circles are classical objects in geometry and physics [3, 4, 18, 186], the latter being preserved by Möbius transformations and thus related to the conformal structure (note that Möbius transformations constitute a special class of conformal transformations characterized by preserving the eigenspaces of the Ricci operator).

A pseudo-Riemannian manifold (M, g) is said to be an *Osserman space* if the eigenvalues of the Jacobi operator are constant on the unit pseudo-sphere bundles. The fact that the local isometries of any locally two-point homogeneous space act transitively on the unit pseudo-sphere bundles shows that any locally two-point homogeneous space is Osserman. The converse is known to be true in Riemannian geometry if dim $M \neq 16$ [83, 206, 208, 209] and it also holds for Lorentzian metrics, even under some weaker assumptions [30, 128]. The situation is, however, quite different when higher signature metrics are considered. Indeed, although some of the two-point homogeneous spaces can be recognized by some Osserman-like properties [32, 44], a remarkable fact is the existence of many non-symmetric and even not locally homogeneous Osserman pseudo-Riemannian metrics. A two-step strategy has been followed so far in the study of Osserman manifolds [157]. We refer to [5, 17] for further details concerning Clifford algebras. The first step in the study of Osserman tensors consists in the determination of the possible algebraic curvature tensors which are Osserman; this is closely related to the existence of certain Clifford structures [44, 83, 206, 208, 209]. The second step is then to classify the manifolds with such a structure. This shows that a fundamental prerequisite for the understanding of Osserman metrics is first to study the problem at a purely algebraic level.

We reiterate the fact that there are many Osserman algebraic curvature tensors which cannot occur in the geometric setting as Osserman manifolds [32, 136, 130, 137], although they can be realized geometrically at a given point. Although the Jacobi operator is probably the most natural operator associated to the curvature tensor, there is some important geometrical information enclosed in some other operators like the Szabó operator (which is defined by the covariant derivative of the curvature tensor), the skew symmetric curvature operator or the higher order Jacobi operator [137]. Moreover, not only the Riemann curvature tensor has been used as a starting object to define curvature operators (cf. [129]). The notion of *conformally Osserman* has been defined analogously using the Weyl conformal curvature tensor if $m \geq 4$ [33, 37]; this notion is conformally invariant. Any Riemannian conformally Osserman manifold is locally conformally equivalent to a rank 1 symmetric space if $m \neq 4, 16$ [33, 210]. Furthermore, any Lorentzian conformally Osserman manifold is locally conformally flat [37].

Some additional related references in both the algebraic and in the geometric context are given by [9, 31, 34, 35, 36, 39, 40, 41, 58, 59, 124, 132, 133, 144, 138, 139, 153, 155, 158, 175, 242, 243, 244].

1.3   ALGEBRAIC PRELIMINARIES

In this section we define certain basic notions. Let V be a real vector space of dimension m. Let $\{e_i\}$ be a basis for V. If $A \in \otimes^4 V^*$, let $A_{ijk\ell} := A(e_i, e_j, e_k, e_\ell)$ be the components of A. Similarly, if $\mathcal{A} \in V^* \otimes V^* \otimes \mathrm{End}(V)$, we shall expand $\mathcal{A}(e_i, e_j)e_k = \mathcal{A}_{ijk}{}^\ell e_\ell$ where we adopt the *Einstein convention* and sum over repeated indices.

1.3.1   JORDAN NORMAL FORM

Let A_a^k be the *Jordan block* of size $k \times k$ for a real number $a \in \mathbb{R}$:

$$
A_a^k := \begin{pmatrix}
a & 1 & 0 & \ldots & 0 & 0 \\
0 & a & 1 & \ldots & 0 & 0 \\
\ldots & \ldots & \ldots & \ldots & \ldots & \ldots \\
0 & 0 & 0 & \ldots & a & 1 \\
0 & 0 & 0 & \ldots & 0 & a
\end{pmatrix}.
$$

Let $\lambda := a + \sqrt{-1}b$ where $b > 0$. We set

$$
A_\lambda := \begin{pmatrix} a & b \\ -b & a \end{pmatrix} \quad \text{and} \quad \mathrm{Id}_2 := \begin{pmatrix} 1 & 0 \\ 0 & 1 \end{pmatrix}.
$$

We define a Jordan block of size $2k \times 2k$ corresponding to the complex eigenvalue λ by setting

$$
A_\lambda^k := \begin{pmatrix}
A_\lambda & \mathrm{Id}_2 & 0 & \ldots & 0 & 0 \\
0 & A_\lambda & \mathrm{Id}_2 & \ldots & 0 & 0 \\
\ldots & \ldots & \ldots & \ldots & \ldots & \ldots \\
0 & 0 & 0 & \ldots & A_\lambda & \mathrm{Id}_2 \\
0 & 0 & 0 & \ldots & 0 & A_\lambda
\end{pmatrix}.
$$

The following is well known, see, for example, [6].

Lemma 1.1.   *Let T be a linear transformation of a real vector space V. Relative to a suitably chosen basis for V, T decomposes as a direct sum of the Jordan blocks described above. Furthermore, the unordered collection of Jordan blocks is uniquely determined by T.*

The *Jordan normal form* of T is the unordered collection of Jordan blocks described above. We say that two linear maps T and $\tilde{T}$ of V are *Jordan equivalent* if any of the following three equivalent conditions are satisfied:

1. There exist bases $\mathcal{B} = \{e_1, \ldots, e_m\}$ and $\tilde{\mathcal{B}} = \{\tilde{e}_1, \ldots, \tilde{e}_m\}$ for V so that the matrix representation of T with respect to the basis $\mathcal{B}$ is equal to the matrix representation of $\tilde{T}$ with respect to the basis $\tilde{\mathcal{B}}$.

2. There exists an isomorphism ψ of V so $T = \psi \tilde{T} \psi^{-1}$; this means that T and $\tilde{T}$ are *conjugate*.

3. The Jordan normal forms of T and $\tilde{T}$ are equal.

We say that T is a *nilpotent operator* if $T^m = 0$ or, equivalently, if 0 is the only eigenvalue of T. We say that $T \neq 0$ is nilpotent of order $r \geq 2$ if $T^r = 0$ but $T^{r-1} \neq 0$; this means that the largest Jordan block of T has size r.

Remark 1.2.   If T is any linear transformation of V, there always exists a non-degenerate inner product $\langle \cdot, \cdot \rangle$ so that T is *self-adjoint* with respect to $\langle \cdot, \cdot \rangle$, i.e., $\langle Tx, y \rangle = \langle x, Ty \rangle$ for all x and y in V. Thus, the condition that T is self-adjoint imposes no constraint on the Jordan normal form of T in the indefinite setting [137].

1.3.2   INDEFINITE GEOMETRY

We fix a non-degenerate inner product $\langle \cdot, \cdot \rangle$ on V; we extend $\langle \cdot, \cdot \rangle$ to an inner product on tensors of all types. A vector v is said to be *spacelike* if $\langle v, v \rangle > 0$, *timelike* if $\langle v, v \rangle < 0$, and *null* if $\langle v, v \rangle = 0$. We say that a subspace W of V is spacelike (resp. timelike or resp. null) if $\langle \cdot, \cdot \rangle$ is positive definite on W (resp. negative definite on W or trivial on W). A null subspace is also often called an *isotropic subspace* or a *degenerate subspace*. The *pseudo-spheres* of unit timelike ($-$) and spacelike ($+$) vectors are defined by setting

$$S^{\pm} = S^{\pm}(V, \langle \cdot, \cdot \rangle) := \{v \in V : \langle v, v \rangle = \pm 1\}. \tag{1.1}$$

We can choose a basis $\{e_i\}$ for V so that

$$\langle e_i, e_j \rangle = \begin{cases} 0 & i \neq j, \\ \pm 1 & i = j. \end{cases}$$

Such a basis is called an *orthonormal basis*. We set $\varepsilon_i := \langle e_i, e_i \rangle$. Let p be the number of indices i with $\varepsilon_i = -1$. Let $q = \dim V - p$ be the complementary index; q is the number of indices i with $\varepsilon_i = +1$. The inner product is then said to have *signature* (p, q); the integers p and q are independent of the particular orthonormal basis chosen. If $\{e_1, \ldots, e_{p+q}\}$ is the standard basis for Euclidean space $\mathbb{R}^m$, then we shall let $\mathbb{R}^{(p,q)}$ be $\mathbb{R}^m$ with the inner product given by:

$$\langle e_i, e_j \rangle = \begin{cases} 0 & i \neq j, \\ -1 & i = j \leq p, \\ 1 & i = j > p. \end{cases}$$

We can construct an isometry between V and $\mathbb{R}^{(p,q)}$ by choosing an orthonormal basis for V of the form given above. However, it is often useful to work in a basis free setting with an abstract vector space V rather than with the concrete realization $\mathbb{R}^{(p,q)}$. The elements of the general linear group which preserve the inner product form a closed subgroup, called the *orthogonal group*, of $\mathrm{GL}(V)$:

$$O(V, \langle \cdot, \cdot \rangle) = \{T \in \mathrm{GL}(V) : T^* \langle \cdot, \cdot \rangle = \langle \cdot, \cdot \rangle\} \, .$$

1.3.3   ALGEBRAIC CURVATURE TENSORS

We say that $A \in \otimes^4 V^*$ is an *algebraic curvature tensor* if A has the symmetries of the Riemann curvature tensor of the Levi-Civita connection (see Section 2.4.1):

$$\begin{aligned} A(x, y, z, w) &= -A(y, x, z, w), \\ A(x, y, z, w) &= A(z, w, x, y), \\ A(x, y, z, w) + A(y, z, x, w) &+ A(z, x, y, w) = 0 \, . \end{aligned} \quad (1.2)$$

We will show presently in Theorem 2.4 that every algebraic curvature tensor is geometrically realizable; thus this is a convenient algebraic context in which to work.

Example 1.3.   If $\phi \in S^2(V^*)$ and $\psi \in \Lambda^2(V^*)$ are symmetric and anti-symmetric bilinear forms, respectively, we may define algebraic curvature tensors by setting:

$$\begin{aligned} A_\phi(x, y, z, w) &:= \phi(x, w)\phi(y, z) - \phi(x, z)\phi(y, w), \\ A_\psi(x, y, z, w) &:= \psi(x, w)\psi(y, z) - \psi(x, z)\psi(y, w) \\ &\quad - 2\psi(x, y)\psi(z, w) \, . \end{aligned} \quad (1.3)$$

If Φ (resp. Ψ) is a self-adjoint (resp. skew adjoint) linear map of V, we may define a corresponding symmetric bilinear form ϕ (resp. alternating bilinear form ψ) by setting $\phi(x, y) := \langle \Phi x, y \rangle$ (resp. $\psi(x, y) := \langle \Psi x, y \rangle$) and thereby obtain an algebraic curvature tensor we shall denote by A_Φ (resp. A_Ψ). Note that $A_\Phi = A_{-\Phi}$ and $A_\Psi = A_{-\Psi}$.

The tensor $\kappa A_{\langle \cdot, \cdot \rangle} = \kappa A_{\mathrm{Id}}$ has constant sectional curvature κ as we shall see shortly; more generally, if L is the second fundamental form of a hypersurface in flat space, then the tensor A_L is the associated curvature tensor of the hypersurface. Thus, tensors of this sort arise geometrically. If J is a Hermitian structure on $(V, \langle \cdot, \cdot \rangle)$, we define the associated Kaehler 2 form

$$\Omega(x, y) := \langle x, Jy \rangle \; ;$$

the tensor $\frac{\kappa}{4}\left(A_{\langle \cdot, \cdot \rangle} + A_\Omega\right) = \frac{\kappa}{4}\{A_{\mathrm{Id}} + A_J\}$ has constant holomorphic sectional curvature κ (cf. [140]). Thus, tensors of this kind also appear naturally in geometry as this is the curvature tensor of the Fubini-Study metric.

One has the following result due to Fiedler [117] – see also [105, 137] for different treatments.

Theorem 1.4. *Let $\Xi(V)$ be the vector space of all algebraic curvature tensors. Then*

$$\Xi(V) = \mathrm{Span}_{\phi \in S^2(V^*)}\{A_\phi\} = \mathrm{Span}_{\psi \in \Lambda^2(V^*)}\{A_\psi\} \, .$$

If $A \in \Xi(V)$ is an algebraic curvature tensor, then $\mathfrak{M} := (V, \langle \cdot, \cdot \rangle, A)$ is said to be a *curvature model*.

1.3.4   HERMITIAN AND PARA-HERMITIAN GEOMETRY

We shall denote the *complex numbers* by $\mathbb{C} := \mathrm{Span}_{\mathbb{R}}\{1, \sqrt{-1}\} = \mathbb{R}^2$. This field is algebraically closed. Similarly, let $\tilde{\mathbb{C}} := \mathrm{Span}_{\mathbb{R}}\{1, P\} = \mathbb{R}^2$ (where $P^2 = 1$) be the *para-complex numbers*. This algebra is not a field but it is closely related to $\mathbb{C}$. Fix a non-degenerate inner product $\langle \cdot, \cdot \rangle$ on V. We say that J is a *pseudo-Hermitian complex structure* on V if J is a linear map of V with

$$J^2 = -\,\mathrm{Id} \quad \text{and} \quad J^*\langle \cdot, \cdot \rangle = \langle \cdot, \cdot \rangle \, .$$

Pseudo-Hermitian structures exist if and only if p and q are both even. We give V the structure of a complex vector space by setting $\sqrt{-1} \cdot v := Jv$. Let $\mathcal{U} = \mathcal{U}(V, \langle \cdot, \cdot \rangle, J)$ be the associated *unitary group*:

$$\mathcal{U} := \{U \in \mathrm{GL}(V) : UJ = JU \quad \text{and} \quad U^*\langle \cdot, \cdot \rangle = \langle \cdot, \cdot \rangle\} \, .$$

If A is an algebraic curvature tensor, the associated *Hermitian curvature model* is the quadruple $\mathfrak{C} := (V, \langle \cdot, \cdot \rangle, J, A)$.

Similarly, a linear map J is said to be a *para-Hermitian structure* on V if

$$J^2 = \mathrm{Id} \quad \text{and} \quad J^*\langle \cdot, \cdot \rangle = -\langle \cdot, \cdot \rangle \, .$$

Such structures exist if and only if $p = q$ so we are in the *neutral signature* setting. The associated *para-unitary group* is given by

$$\tilde{\mathcal{U}} := \{U \in \mathrm{GL}(V) : UJ = JU \quad \text{and} \quad U^*\langle \cdot, \cdot \rangle = \langle \cdot, \cdot \rangle\} \, .$$

If A is an algebraic curvature tensor, the associated *para-Hermitian curvature model* is the quadruple $\tilde{\mathfrak{C}} := (V, \langle \cdot, \cdot \rangle, J, A)$.

Let $(V, \langle \cdot, \cdot \rangle, J, A)$ be a Hermitian curvature model (resp. a para-Hermitian curvature model). Let x be a non-null vector and $\pi := \mathrm{Span}\{x, Jx\}$ be the corresponding non-degenerate *holomorphic* (resp. *para-holomorphic*) 2-plane. Let $\varepsilon = +1$ in the Hermitian setting and $\varepsilon = -1$ in the para-Hermitian setting. The *holomorphic sectional curvature* (resp. *para-holomorphic sectional curvature*) is defined to be

$$K(\pi) := \varepsilon A(x, Jx, Jx, x) \langle x, x \rangle^{-2} .$$

The model is said to have constant holomorphic sectional curvature (resp. constant para-holomorphic sectional curvature) if $K(\pi)$ is constant for all non-degenerate holomorphic planes π. The following curvature tensor has constant holomorphic (resp. para holomorphic) sectional curvature:

$$A := a A_{\mathrm{Id}} + b A_J \quad \text{for} \quad a, b \in \mathbb{R} .$$

There are two other algebras which will be important in our development. Let

$$\mathbb{H} := \mathrm{Span}_{\mathbb{R}}\{1, J_1, J_2, J_3\} = \mathbb{R}^4$$

be the *quaternions* where the algebra structure is given by the Clifford commutation relations :

$$J_1^2 = J_2^2 = J_3^2 = -\mathrm{Id}, \quad J_1 J_2 = -J_2 J_1 = J_3, \quad J_2 J_3 = -J_3 J_2 = J_1, \quad J_3 J_1 = -J_1 J_3 = J_2 .$$

This is a division algebra which is not commutative. We call $\{J_1, J_2, J_3\}$ a *hyper-Hermitian structure* if we have the compatibility relations between $\{J_1, J_2, J_3\}$ and $(V, \langle \cdot, \cdot \rangle)$:

$$J_1^* \langle \cdot, \cdot \rangle = J_2^* \langle \cdot, \cdot \rangle = J_3^* \langle \cdot, \cdot \rangle = \langle \cdot, \cdot \rangle .$$

Finally, we let the *para-quaternions* be defined by

$$\widetilde{\mathbb{H}} := \mathrm{Span}_{\mathbb{R}}\{1, J_1, J_2, J_3\} = \mathbb{R}^4$$

where the algebra structure is given by the relations:

$$J_1^2 = -\mathrm{Id}, \quad J_2^2 = J_3^2 = \mathrm{Id},$$
$$J_1 J_2 = -J_2 J_1 = J_3, \quad J_2 J_3 = -J_3 J_2 = J_1, \quad J_3 J_1 = -J_1 J_3 = J_2 .$$

This algebra is neither a division algebra nor is it commutative. Still it has many of the features of $\mathbb{H}$. We call $\{J_1, J_2, J_3\}$ a *hyper-para-Hermitian structure* if we have the following compatibility relations:

$$J_1^* \langle \cdot, \cdot \rangle = \langle \cdot, \cdot \rangle \quad \text{and} \quad J_2^* \langle \cdot, \cdot \rangle = J_3^* \langle \cdot, \cdot \rangle = -\langle \cdot, \cdot \rangle .$$

1.3.5   THE JACOBI AND SKEW SYMMETRIC CURVATURE OPERATORS

Let $\mathfrak{M} := (V, \langle \cdot, \cdot \rangle, A)$ be a curvature model. The associated Jacobi operator $\mathcal{J} = \mathcal{J}_{\mathfrak{M}}$ and skew symmetric curvature operator $\mathcal{A} = \mathcal{A}_{\mathfrak{M}}$ are characterized, respectively, by the identities:

$$\langle \mathcal{J}(x)y, z \rangle = A(y, x, x, z) \quad \text{and} \quad \langle \mathcal{A}(x, y)z, w \rangle = A(x, y, z, w). \tag{1.4}$$

The following is a useful observation:

Lemma 1.5.   *If k is odd, then* $\mathrm{Tr}\{\mathcal{A}(x, y)^k\} = 0$.

Proof.   If $k = 1$, we use the curvature symmetries to see $A(x, y, e_i, e_j) = -A(x, y, e_j, e_i)$ so:

$$\mathrm{Tr}\{A(x, y)\} = \sum_i \varepsilon_i A(x, y, e_i, e_i) = 0 \ .$$

Similarly, if $k = 3$, we compute:

$$
\begin{aligned}
\mathrm{Tr}\{A(x, y)^3\} &= \sum_{i,j,k} \varepsilon_i \varepsilon_j \varepsilon_k A(x, y, e_i, e_j) A(x, y, e_j, e_k) A(x, y, e_k, e_i) \\
&= -\sum_{i,j,k} \varepsilon_i \varepsilon_j \varepsilon_k A(x, y, e_j, e_i) A(x, y, e_k, e_j) A(x, y, e_i, e_k) \\
&= -\sum_{i,j,k} \varepsilon_i \varepsilon_j \varepsilon_k A(x, y, e_j, e_i) A(x, y, e_i, e_k) A(x, y, e_k, e_j) \\
&= -\mathrm{Tr}\{A(x, y)^3\}
\end{aligned}
$$

so $\mathrm{Tr}\{A(x, y)^3\} = 0$. The proof for general odd k is similar and is therefore omitted. $\square$

1.3.6   SECTIONAL, RICCI, SCALAR, AND WEYL CURVATURE

Let $\mathfrak{M} := (V, \langle \cdot, \cdot \rangle, A)$ be a curvature model. The sectional curvature $K = K_{\mathfrak{M}}$ of a non-degenerate oriented 2-plane $\pi = \mathrm{Span}\{x, y\}$ is given by setting:

$$K(\pi) = \frac{A(y, x, x, y)}{\langle x, x \rangle \langle y, y \rangle - \langle x, y \rangle^2}. \tag{1.5}$$

This is independent of the particular oriented basis which is chosen.

Remark 1.6.   We adopt the notation of Example 1.3 to define $A_{\langle \cdot, \cdot \rangle} = A_{\mathrm{Id}}$. Suppose that one is given a curvature model $\mathfrak{M} := (V, \langle \cdot, \cdot \rangle, A)$. Then $K_{\mathfrak{M}}(\pi) = \kappa$ for all π if and only if $A = \kappa A_{\langle \cdot, \cdot \rangle}$; these are the tensors of *constant sectional curvature*.

The *Ricci tensor* is the symmetric 2 tensor defined by setting:

$$\rho(x, y) := \mathrm{Tr}\{z \to \mathcal{A}(z, x)y\}; \qquad \rho(x, x) = \mathrm{Tr}\{\mathcal{J}(x)\}. \tag{1.6}$$

By an abuse of notation, we shall also let ρ denote the Ricci operator characterized by the identity $\langle \rho x, y \rangle = \rho(x, y)$. Let $\varepsilon_{ij} := \langle e_i, e_j \rangle$ be the components of the inner product; let ε^{ij} denote the inverse matrix. Then $\rho_{ij} := \varepsilon^{k\ell} A_{ik\ell j}$. The *scalar curvature* is defined by a final contraction of indices:

$$\tau := \varepsilon^{ij} \rho_{ij} = \varepsilon^{ij} \varepsilon^{k\ell} A_{ik\ell j} = \mathrm{Tr}_{\langle \cdot, \cdot \rangle}\{\rho\} . \tag{1.7}$$

We say that $\mathfrak{M}$ is *Einstein* if there is a constant c so that $\rho = c \langle \cdot, \cdot \rangle$. The scalar curvature $\tau = mc$ in this setting so $\rho = \frac{\tau}{m} \langle \cdot, \cdot \rangle$.

Let $\mathfrak{C} = (V, \langle \cdot, \cdot \rangle, J, A)$ be a pseudo-Hermitian curvature model. We define the Kaehler form $\Omega \in \Lambda^2(V^*)$, the $\star$-Ricci tensor $\rho^\star \in \otimes^2 V^*$ and the $\star$-scalar curvature $\tau^\star \in \mathbb{R}$ by setting

$$\Omega(x, y) := \langle x, Jy \rangle, \quad \rho^\star_{ij} := \varepsilon^{k\ell} A(e_k, e_i, Je_j, Je_\ell), \quad \tau^\star := \varepsilon^{ij} \rho^\star_{ij} = \mathrm{Tr}_{\langle \cdot, \cdot \rangle}\{\rho^\star\} . \tag{1.8}$$

Similarly, if $\tilde{\mathfrak{C}} = (V, \langle \cdot, \cdot \rangle, J, A)$ is a para-Hermitian curvature model, we define

$$\Omega(x, y) := \langle x, Jy \rangle, \quad \rho^\star_{ij} := -\varepsilon^{k\ell} A(e_k, e_i, Je_j, Je_\ell), \quad \tau^\star := \varepsilon^{ij} \rho^\star_{ij} = \mathrm{Tr}_{\langle \cdot, \cdot \rangle}\{\rho^\star\} . \tag{1.9}$$

Note that ρ is always a symmetric tensor field. However, $\rho^\star$ need not be symmetric. We say that a Hermitian model $\mathfrak{C}$ or a para-Hermitian model $\tilde{\mathfrak{C}}$ is $\star$-*Einstein* if $\rho^\star = \frac{\tau^\star}{m} \langle \cdot, \cdot \rangle$.

Let $\mathfrak{M}$ be a curvature model. We define the associated *Weyl conformal curvature tensor* W and the *Schouten tensor* C by setting:

$$W(x, y, z, w) := A(x, y, z, w) + \frac{\tau}{(m-1)(m-2)}\{\langle x, w \rangle \langle y, z \rangle - \langle x, z \rangle \langle y, w \rangle\} \tag{1.10}$$
$$- \frac{1}{m-2}\{\langle x, w \rangle \rho(y, z) - \rho(x, z) \langle y, w \rangle + \rho(x, w) \langle y, z \rangle - \langle x, z \rangle \rho(y, w)\},$$
$$C(x, y) := \frac{1}{m-2}\{\rho(x, y) - \frac{\tau}{2(m-1)} \langle x, y \rangle\} . \tag{1.11}$$

Note that these tensors are related by the fact that $\rho = 0$ if and only if $C = 0$, which occurs if and only if $A = W$.

1.3.7   CURVATURE DECOMPOSITIONS

The Ricci tensor defines an $O(V, \langle \cdot, \cdot \rangle)$ equivariant short exact sequence:

$$0 \to \ker(\rho) \to \Xi(V) \to S^2(V^*) \to 0$$

which is equivariantly split by the map $A \to W_A$. The space of all Weyl conformal curvature tensors is defined to be

$$\ker(\rho) = \mathrm{Span}_{A \in \Xi(V)}\{W_A\} .$$

Let $S^2_0(V^*)$ be the set of trace free symmetric 2 tensors; $S^2(V^*) = S^2_0(V^*) \oplus \mathbb{R} \cdot \langle \cdot, \cdot \rangle$. One has the following result due to Singer and Thorpe [241]:

Theorem 1.7. *We have an orthogonal direct sum decomposition of $\Xi(V)$ into irreducible $O(V, \langle \cdot, \cdot \rangle)$ modules $\Xi(V) = \ker(\rho) \oplus S^2_0(V^*) \oplus \mathbb{R}$.*

Tricerri and Vanhecke [251] constructed a similar decomposition of $\Xi(V)$ in the Hermitian setting. Let $\langle \cdot, \cdot \rangle$ be a positive definite inner product on V and let J be a Hermitian complex structure. One may decompose $V^* \otimes V^*$ as the direct sum of 6 irreducible $\mathcal{U}$ modules:

$$V^* \otimes V^* = \langle \cdot, \cdot \rangle \cdot \mathbb{R} \oplus S_{0,+}^2(V^*) \oplus S_-^2(V^*) \oplus \Omega \cdot \mathbb{R} \oplus \Lambda_{0,+}^2(V^*) \oplus \Lambda_-^2(V^*) \text{ where}$$

$$S_{0,+}^2(V^*) := \{\theta \in S^2(V^*) : J^*\theta = \theta \quad \text{and} \quad \theta \perp \langle \cdot, \cdot \rangle \},$$

$$S_-^2(V^*) := \{\theta \in S^2(V^*) : J^*\theta = -\theta \},$$

$$\Lambda_{0,+}^2(V^*) := \{\theta \in \Lambda^2(V^*) : J^*\theta = \theta \quad \text{and} \quad \theta \perp \Omega \},$$

$$\Lambda_-^2(V^*) := \{\theta \in \Lambda^2(V^*) : J^*\theta = -\theta \}.$$

Theorem 1.8. *Let $(\langle \cdot, \cdot \rangle, J)$ be a Hermitian structure on V.*

1. *We have an orthogonal direct sum decomposition of $\Xi(V)$ into irreducible $\mathcal{U}$ modules:*

   (a) *If $2n = 4$, $\Xi(V) = \mathcal{W}_1 \oplus \mathcal{W}_2 \oplus \mathcal{W}_3 \oplus \mathcal{W}_4 \oplus \mathcal{W}_7 \oplus \mathcal{W}_8 \oplus \mathcal{W}_9$.*

   (b) *If $2n = 6$, $\Xi(V) = \mathcal{W}_1 \oplus \mathcal{W}_2 \oplus \mathcal{W}_3 \oplus \mathcal{W}_4 \oplus \mathcal{W}_5 \oplus \mathcal{W}_7 \oplus \mathcal{W}_8 \oplus \mathcal{W}_9 \oplus \mathcal{W}_{10}$.*

   (c) *If $2n \geq 8$, $\Xi(V) = \mathcal{W}_1 \oplus \mathcal{W}_2 \oplus \mathcal{W}_3 \oplus \mathcal{W}_4 \oplus \mathcal{W}_5 \oplus \mathcal{W}_6 \oplus \mathcal{W}_7 \oplus \mathcal{W}_8 \oplus \mathcal{W}_9 \oplus \mathcal{W}_{10}$.*

   $\mathcal{W}_1 \approx \mathcal{W}_4$ and, if $2n \geq 6$, $\mathcal{W}_2 \approx \mathcal{W}_5$. The other $\mathcal{U}$ modules appear with multiplicity 1.

2. *We have that:*

   (a) $\tau \oplus \tau^\star : \mathcal{W}_1 \oplus \mathcal{W}_4 \approx \mathbb{R} \oplus \mathbb{R}$.

   (b) *If $2n = 4$, $\rho_{0,+,s} : \mathcal{W}_2 \approx S_{0,+}^2(V^*)$.*

   (c) *If $2n \geq 6$, $\rho_{0,+,s} \oplus \rho_{0,+,s}^\star : \mathcal{W}_2 \oplus \mathcal{W}_5 \approx S_{0,+}^2(V^*) \oplus S_{0,+}^2(V^*)$.*

   (d) $\mathcal{W}_3 = \{A \in \Xi(V) : A(x, y, z, w) = A(Jx, Jy, z, w) \ \forall x, y, z, w\} \cap \ker(\rho)$.

   (e) *If $2n \geq 8$, $\mathcal{W}_6 = \ker(\rho \oplus \rho^\star) \cap \{A \in \Xi(V) : J^*A = A\} \cap \mathcal{W}_3^\perp$.*

   (f) $\mathcal{W}_7 = \{A \in \Xi(V) : A(Jx, y, z, w) = A(x, y, Jz, w) \ \forall x, y, z, w\}$.

   (g) $\rho_{-,s} : \mathcal{W}_8 \approx S_-^2(V^*)$.

   (h) $\rho_{-,\Lambda}^\star : \mathcal{W}_9 \approx \Lambda_-^2(V^*)$.

   (i) *If $2n \geq 6$, $\mathcal{W}_{10} = \{A \in \Xi(V) : J^*A = -A\} \cap \ker(\rho \oplus \rho^\star)$.*

3. *Let $m = \dim(V) = 2n$. The dimensions of these modules are given by:*

	m = 4	m = 6	m ≥ 8		m = 4	m = 6	m ≥ 8
$\mathcal{W}_1$	1	1	1	$\mathcal{W}_2$	3	8	$n^2 - 1$
$\mathcal{W}_3$	5	27	$\frac{n^2(n-1)(n+3)}{4}$	$\mathcal{W}_4$	1	1	1
$\mathcal{W}_6$	0	0	$\frac{n^2(n+1)(n-3)}{4}$	$\mathcal{W}_5$	0	8	$n^2 - 1$
$\mathcal{W}_7$	2	12	$\frac{n^2(n^2-1)}{6}$	$\mathcal{W}_8$	6	12	$n^2 + n$
$\mathcal{W}_{10}$	0	30	$\frac{2n^2(n^2-4)}{3}$	$\mathcal{W}_9$	2	6	$n^2 - n$

Remark 1.9. There are other related curvature decompositions. The assumption that the inner product is positive definite is an inessential one; there are pseudo-Hermitian curvature decompositions. With appropriate changes of sign, there is an analogous para-Hermitian curvature decomposition [60]. Additionally, there is also a hyper-Hermitian curvature decomposition [96]. The factor $\mathcal{W}_7$ plays a special role in the analysis; tensors in $\mathcal{W}_7^\perp$ are said to satisfy the *Gray identity* and are geometrically realizable by Hermitian manifolds as we shall see subsequently in Section 2.5.

1.3.8   SELF-DUALITY AND ANTI-SELF-DUALITY CONDITIONS

If $m = 4$, $\ker(\rho)$ is not irreducible as an $SO(V, \langle \cdot, \cdot \rangle)$ module. We restrict to signature $(2, 2)$ as the analysis in other signatures is similar. Let $\star : \Lambda^2(V^*) \to \Lambda^2(V^*)$ be the *Hodge* operator. If *orn* is the oriented volume form, the Hodge operator is characterized by the identity:

$$\langle \omega_1, \omega_2 \rangle orn = \omega_1 \wedge \star \omega_2$$

for $\omega_i \in \Lambda^2(V^*)$. Since $\star^2 = \mathrm{Id}$, we may decompose $\Lambda^2(V^*)$ into the ± 1 eigenspaces of $\star$:

$$\Lambda^2(V^*) = \Lambda^+(V^*) \oplus \Lambda^-(V^*) \ .$$

Let $\{e_1, e_2, e_3, e_4\}$ be an orthonormal basis for V where e_1 and e_2 are spacelike while e_3 and e_4 are timelike vectors. Set

$$
\begin{aligned}
E_1^\pm &= \tfrac{1}{\sqrt{2}}\{e^1 \wedge e^2 \pm e^3 \wedge e^4\}, \quad E_2^\pm = \tfrac{1}{\sqrt{2}}\{e^1 \wedge e^3 \pm e^2 \wedge e^4\}, \\
E_3^\pm &= \tfrac{1}{\sqrt{2}}\{e^1 \wedge e^4 \mp e^2 \wedge e^3\}.
\end{aligned}
\tag{1.12}
$$

Then $\{E_1^\pm \, E_2^\pm, E_3^\pm\}$ is an orthonormal basis for $\Lambda^\pm$. We also note that

$$\langle E_1^\pm, E_1^\pm \rangle = 1, \quad \langle E_2^\pm, E_2^\pm \rangle = -1, \quad \langle E_3^\pm, E_3^\pm \rangle = -1 \ .$$

Let $\mathfrak{M}$ be a curvature model. Since $A(\cdot, \cdot, \cdot, \cdot)$ is anti-symmetric in the first pair of indices and in the second pair of indices, we may regard $A : \Lambda^2 \otimes \Lambda^2 \to \mathbb{R}$. A similar remark pertains to the associated Weyl conformal curvature tensor defined in Equation (1.10). The Weyl operator W induces operators $W^\pm : \Lambda^\pm \longrightarrow \Lambda^\pm$. These operators have the following matrix form with respect to the bases defined above:

$$
W^\pm = \begin{pmatrix}
W_{11}^\pm & W_{12}^\pm & W_{13}^\pm \\
-W_{12}^\pm & -W_{22}^\pm & -W_{23}^\pm \\
-W_{13}^\pm & -W_{23}^\pm & -W_{33}^\pm
\end{pmatrix},
\tag{1.13}
$$

where $W_{ij}^\pm = W(E_i^\pm, E_j^\pm)$ and $W(e^i \wedge e^j, e^k \wedge e^\ell) = W(e_i, e_j, e_k, e_\ell)$.

Definition 1.10. A curvature model $\mathfrak{M} := (V, \langle \cdot, \cdot \rangle, A)$ is called *self-dual* (resp. *anti-self-dual*) if $W^- = 0$ (resp. $W^+ = 0$).

Let ρ_0 denote the trace free Ricci tensor. We extend ρ_0 and A to maps from $\Lambda^2(V^*)$ to $\Lambda^2(V^*)$. We then have a decomposition

$$A = \frac{\tau}{12} \operatorname{Id}_{\Lambda^2} + \rho_0 + \begin{pmatrix} W^+ & 0 \\ 0 & W^- \end{pmatrix} . \tag{1.14}$$

This decomposition will play a central role in our discussion of Osserman curvature models subsequently in Theorem 1.21.

1.4   SPECTRAL GEOMETRY OF THE CURVATURE OPERATOR

In this section, we examine spectral properties of the curvature operator and various commutativity properties. Throughout this section, we shall fix a curvature model $\mathfrak{M} = (V, \langle \cdot, \cdot \rangle, A)$.

If T is a linear transformation of V, the characteristic polynomial of T is:

$$p_\lambda(T) := \det(T - \lambda \operatorname{Id}) .$$

A complex number ν is said to be an *eigenvalue* of T or, equivalently, is said to be in the *spectrum of* T if and only if $p_\nu(T) = 0$, i.e., if $T - \nu \operatorname{Id}$ is not invertible or, equivalently, if T has a non-zero complex eigenvector with eigenvalue ν. Let $\operatorname{Spec}\{T\}$ be the collection of complex eigenvalues of T where each eigenvalue is repeated according to its multiplicity as a root of $p_\lambda(T)$. The coefficients of λ in $p_\lambda(T)$ are the elementary symmetric functions of the eigenvalues. Let $M_m(\mathbb{C})$ be the set of $m \times m$ complex matrices. The following is well known:

Lemma 1.11.   *Let $T_i \in M_m(\mathbb{C})$. The following conditions are equivalent:*

1. $\operatorname{Spec}\{T_1\} = \operatorname{Spec}\{T_2\}$.

2. $\operatorname{Tr}\{T_1^i\} = \operatorname{Tr}\{T_2^i\}$ *for* $1 \le i \le m$.

3. $p_\lambda(T_1) = p_\lambda(T_2)$.

The following is a useful observation which also is well known:

Lemma 1.12.   *Let p be a polynomial function on V. If p vanishes on an open subset $\mathcal{O}$ of V, then p vanishes identically.*

1.4.1   OSSERMAN AND CONFORMALLY OSSERMAN MODELS

Let $\mathcal{J}(x)$ be the Jacobi operator which was introduced in Section 1.3.5. We say that the curvature model $\mathfrak{M}$ is a *spacelike* (resp. *timelike*) Osserman curvature model if the eigenvalues of $\mathcal{J}$ are constant on the unit pseudo-spheres $S^\pm$ of unit spacelike (+) and unit timelike (−) vectors as defined in Equation (1.1). One says $\mathfrak{M}$ is a *null Osserman* curvature model if $\mathcal{J}(x)$ is a nilpotent operator for every null vector x. The adjective "Osserman" is used following the seminal paper by Osserman [218].

Assume $\mathfrak{M}$ is a model of signature (p, q). The conditions spacelike Osserman and timelike Osserman are equivalent if $p > 0$ and $q > 0$ so the pseudo-spheres S^+ and S^- are both non-empty [129, 137]:

Theorem 1.13.   *Let $\mathfrak{M}$ be a curvature model of signature (p, q) with $p, q > 0$.*

1. *The following conditions are equivalent and if either is satisfied, then $\mathfrak{M}$ is said to be an* Osserman *curvature model:*

   (a) $\mathfrak{M}$ is a timelike Osserman curvature model.

   (b) $\mathfrak{M}$ is a spacelike Osserman curvature model.

   (c) There exist constants c_i so $\mathrm{Tr}\{\mathcal{J}(x)^i\} = c_i \langle x, x \rangle^i$ for any $x \in V, 1 \le i \le m$.

2. *If $\mathfrak{M}$ is an Osserman curvature model, then $\mathfrak{M}$ is null Osserman and Einstein.*

Remark 1.14.   One has that $\mathfrak{M}$ is *Einstein* (1-stein) if $\mathrm{Tr}\{\mathcal{J}(x)\} = c_1 \langle x, x \rangle$ for all x. More generally, one says that $\mathfrak{M}$ is *k-stein* if $\mathrm{Tr}\{\mathcal{J}(x)^i\} = c_i \langle x, x \rangle^i$ for $1 \le i \le k$ and all x. One then has $\mathfrak{M}$ is Osserman if and only if $\mathfrak{M}$ is m-stein. In the Lorentzian setting, a stronger result holds. One has that a Lorentzian curvature model is Osserman if and only if it is 2-stein; we refer to [30] for details.

Remark 1.15.   If $\mathfrak{M}$ is Osserman, the characteristic polynomial has the form

$$p_\lambda(\mathcal{J}(x)) = a_0 \pm a_1 \lambda + a_2 \lambda^2 \pm a_3 \lambda^3 \ldots \quad \text{for} \quad x \in S^\pm$$

and suitably chosen coefficients a_i. Thus, the eigenvalues of the Jacobi operator can be different on S^+ and S^-, although, of course, they remain constant on each of S^- and S^+.

If the inner product is positive definite, then $\mathcal{J}(x)$ is necessarily diagonalizable and thus the eigenvalue structure determines the Jordan normal form. This is not the case, however, in the indefinite setting. A curvature model $\mathfrak{M}$ is called *Jordan Osserman* if the Jordan normal form of the

Jacobi operators is constant on the pseudo-spheres $S^{\pm}$. Clearly, Jordan Osserman implies Osserman, but the converse is not true in general as we shall see in Example 3.10.

The structure of a Jordan Osserman algebraic curvature tensor strongly depends on the signature (p, q) of the metric tensor. One has, for example, the result [142]:

Theorem 1.16. *Let $\mathfrak{M} = (V, \langle \cdot, \cdot \rangle, A)$ be a curvature model. If $p < q$ and if $\mathfrak{M}$ is spacelike Jordan Osserman, then $\mathcal{J}(x)$ is diagonalizable for $x \in S^+$.*

On the other hand, the Jordan normal form can be arbitrarily complicated in neutral signature $p = q$. We refer to [141] for the following result:

Theorem 1.17. *Let $J_0 \in M_\ell(\mathbb{R})$. There exists $v = v(\ell)$ and an algebraic curvature tensor A on $\mathbb{R}^{(v,v)}$ so that $\mathcal{J}(x)$ is conjugate to $\pm J_0 \oplus 0$ for $x \in S^{\pm}$.*

If $\mathfrak{M} = (V, \langle \cdot, \cdot \rangle, A)$ is a curvature model, then the associated Weyl tensor again satisfies the symmetries of Equation (1.2). The *conformal Jacobi operator* $\mathcal{J}_W$ is then defined by W. One says that $\mathfrak{M}$ is a *conformally Osserman* curvature model if $(V, \langle \cdot, \cdot \rangle, W)$ is an Osserman curvature model, i.e., if the eigenvalues of $\mathcal{J}_W$ are constant on the pseudo-spheres S^+ and S^-.

1.4.2   OSSERMAN CURVATURE MODELS IN SIGNATURE $(2, 2)$

In Chapter 6 we shall pay special attention to the geometry of Walker manifolds in dimension 4, focusing in particular on spectral properties of the Jacobi operator. We refer to [32, 130] for the proof of the following classification result:

Theorem 1.18. *Let $\mathfrak{M}$ be a curvature model of signature $(2, 2)$. Then $\mathfrak{M}$ is Osserman if and only if one of the following holds for every $x \in S^+$:*

1. *Type Ia: the Jacobi operators are diagonalizable*

$$\mathcal{J}(x) = \begin{pmatrix} \alpha & 0 & 0 \\ 0 & \beta & 0 \\ 0 & 0 & \gamma \end{pmatrix},$$

   which occurs if and only if there exists an orthonormal basis $\{e_1, e_2, e_3, e_4\}$ for V where the non-zero components of A are given by

$$\begin{array}{ll} A_{1221} = A_{4334} = \alpha, & A_{1331} = A_{4224} = -\beta, \\ A_{1441} = A_{3223} = -\gamma, & A_{1234} = (-2\alpha + \beta + \gamma)/3, \\ A_{1423} = (\alpha + \beta - 2\gamma)/3, & A_{1342} = (\alpha - 2\beta + \gamma)/3. \end{array}$$

2. *Type Ib: there is a complex eigenvalue for the Jacobi operators*

$$\mathcal{J}(x) = \begin{pmatrix} \alpha & \beta & 0 \\ -\beta & \alpha & 0 \\ 0 & 0 & \gamma \end{pmatrix},$$

which occurs if and only if there exists an orthonormal basis $\{e_1, e_2, e_3, e_4\}$ for V where the non-zero components of A are given by

$$\begin{aligned}
A_{1221} &= A_{4334} = \alpha, & A_{1331} &= A_{4224} = -\alpha, \\
A_{1441} &= A_{3223} = -\gamma, & A_{2113} &= A_{2443} = -\beta, \\
A_{1224} &= A_{1334} = \beta, & A_{1234} &= (-\alpha + \gamma)/3, \\
A_{1423} &= 2(\alpha - \gamma)/3, & A_{1342} &= (-\alpha + \gamma)/3.
\end{aligned}$$

3. *Type II: the Jacobi operators can be written in the form*

$$\mathcal{J}(x) = \begin{pmatrix} \pm(\alpha - \tfrac{1}{2}) & \pm\tfrac{1}{2} & 0 \\ \mp\tfrac{1}{2} & \pm(\alpha + \tfrac{1}{2}) & 0 \\ 0 & 0 & \beta \end{pmatrix},$$

which occurs if and only if there exists an orthonormal basis $\{e_1, e_2, e_3, e_4\}$ for V where the non-zero components of A are given by

$$A_{1221} = A_{4334} = \pm\left(\alpha - \tfrac{1}{2}\right), \quad A_{1331} = A_{4224} = \mp\left(\alpha + \tfrac{1}{2}\right),$$

$$A_{1441} = A_{3223} = -\beta, \quad A_{2113} = A_{2443} = \mp\tfrac{1}{2},$$

$$A_{1224} = A_{1334} = \pm\tfrac{1}{2}, \quad A_{1234} = \left(\pm\left(-\alpha + \tfrac{3}{2}\right) + \beta\right)/3,$$

$$A_{1423} = 2(\pm\alpha - \beta)/3, \quad A_{1342} = \left(\pm\left(-\alpha - \tfrac{3}{2}\right) + \beta\right)/3.$$

4. *Type III: the Jacobi operators can be written in the form*

$$\mathcal{J}(x) = \begin{pmatrix} \alpha & 0 & \tfrac{\sqrt{2}}{2} \\ 0 & \alpha & \tfrac{\sqrt{2}}{2} \\ -\tfrac{\sqrt{2}}{2} & \tfrac{\sqrt{2}}{2} & \alpha \end{pmatrix},$$

which occurs if and only if there exists an orthonormal basis $\{e_1, e_2, e_3, e_4\}$ for V where the non-zero components of A are given by

$$A_{1221} = A_{4334} = \alpha, \quad A_{1331} = A_{4224} = -\alpha, \quad A_{1441} = A_{3223} = -\alpha,$$

$$A_{2114} = A_{2334} = -\sqrt{2}/2, \quad A_{3114} = -A_{3224} = \sqrt{2}/2,$$

$$A_{1223} = A_{1443} = A_{1332} = -A_{1442} = \sqrt{2}/2.$$

Remark 1.19. Types II and III correspond to the existence of a double and triple root for the minimal polynomial of $\mathcal{J}(x)$, respectively. Type Ib corresponds to a complex root of the minimal polynomial. Theorem 1.18 shows that the Jordan normal form completely characterizes a Jordan Osserman curvature model of signature $(2, 2)$ up to isomorphism. Furthermore, Osserman and Jordan Osserman are equivalent concepts in signature $(2, 2)$; this fails in signature $(3, 3)$ or higher as Example 3.10 illustrates.

We follow the discussion in [37] to give another description of Osserman curvature models of signature $(2, 2)$. Consider a hyper-para-Hermitian structure $\{J_1, J_2, J_3\}$ as defined in Section 1.3.4. The hyper-para-Hermitian structure gives rise to three new natural operators: two skew adjoint 2-step nilpotent operators given by $\Phi_1 - \Phi_2$ and $\Phi_1 - \Phi_3$ and one para-Hermitian structure $(\Phi_2 - \Phi_3)/\sqrt{2}$. We adopt the notation established in Example 1.3. One obtains a complete description of Osserman signature $(2, 2)$ curvature models [37]:

Theorem 1.20. *A curvature model $\mathfrak{M}$ of signature $(2, 2)$ is Osserman if and only if there exists a hyper-para-Hermitian structure $\{J_1, J_2, J_3\}$ and constants λ_i, λ_{ij} such that*

$$A = \lambda_0 A_{\langle \cdot, \cdot \rangle} + \sum_i \lambda_i A_{J_i} + \sum_{i<j} \lambda_{ij} \left(A_{J_i} + A_{J_j} - A_{J_i - J_j} \right).$$

In Section 1.3.8, we introduced the Hodge $\star$ operator and decomposed the Weyl conformal curvature tensor in the form $W = W^+ \oplus W^-$. We have the following characterization [130]:

Theorem 1.21. *Let $\mathfrak{M} = (V, \langle \cdot, \cdot \rangle, A)$ be an Einstein curvature model of signature $(2, 2)$. Then $\mathfrak{M}$ is Osserman if and only if $\mathfrak{M}$ is either self-dual or anti-self-dual.*

Proof. Recall the decomposition of Equation (1.14) and the notation of Equation (1.12). We also recall Equation (1.13). If A is an Osserman algebraic curvature tensor, then specialize the orthonormal basis $\{e_i\}$ of Theorem 1.18 to get after a straightforward calculation that W^- vanishes and moreover the Jordan normal form of the self-dual part of the Weyl tensor corresponds to the Jordan normal form of the Jacobi operators $\mathcal{J}$. A change of orientation on the basis would give the analogous anti-self-dual result. Conversely, if A is assumed to be self-dual and $\rho_A = \frac{1}{4}\tau_A \langle \cdot, \cdot \rangle$, then Equation (1.14) becomes

$$A = \frac{\tau_A}{12} \operatorname{Id}_{\Lambda^2} + \begin{pmatrix} W^+ & 0 \\ 0 & 0 \end{pmatrix} \tag{1.15}$$

from where it follows that A is Osserman proceeding as in [8, 130, 157]. □

1.4.3   IVANOV–PETROVA CURVATURE MODELS

Let $(p, q) \in \{(2, 0), (1, 1), (0, 2)\}$. Let $\mathrm{Gr}^+_{(p,q)}$ be the *Grassmannian* of oriented 2-planes of signature (p, q). Let $\mathrm{Gr}^+_2 = \mathrm{Gr}_{(2,0)} \sqcup \mathrm{Gr}_{(1,1)} \sqcup \mathrm{Gr}_{(0,2)}$ be the Grassmannian of oriented non-degenerate 2-planes. Let

$$\mathcal{O}_{(2,0)} := \{(x, y) \in V \oplus V : \langle x, x \rangle \langle y, y \rangle - \langle x, y \rangle^2 > 0 \text{ and } \langle x, x \rangle < 0\},$$
$$\mathcal{O}_{(1,1)} := \{(x, y) \in V \oplus V : \langle x, x \rangle \langle y, y \rangle - \langle x, y \rangle^2 < 0\},$$
$$\mathcal{O}_{(0,2)} := \{(x, y) \in V \oplus V : \langle x, x \rangle \langle y, y \rangle - \langle x, y \rangle^2 > 0 \text{ and } \langle x, x \rangle > 0\}.$$

Then $\pi := \mathrm{Span}\{x, y\} \in \mathrm{Gr}^+_{(p,q)}$ if and only if $(x, y) \in \mathcal{O}_{(p,q)}$. For such a π, we define the *skew symmetric curvature operator* $\mathcal{A}(\pi)$ by setting

$$\mathcal{A}(\pi) := \left| \langle x, x \rangle \langle y, y \rangle - \langle x, y \rangle^2 \right|^{-1/2} \mathcal{A}(x, y).$$

This operator is independent of the particular oriented basis $\{x, y\}$ which is chosen. One says that $\mathfrak{M}$ is a spacelike, timelike or mixed *Ivanov–Petrova* curvature model if the eigenvalues of $\mathcal{A}(\pi)$ are constant on the Grassmannian of all oriented non-degenerate 2-planes of the appropriate signature. The notation is chosen owing to the seminal papers by Ivanov and Petrova [171, 172]. Theorem 1.13 generalizes to this setting to yield the following result [137]:

Theorem 1.22.   *Let $\mathfrak{M}$ be a curvature model of signature (p, q) with $p, q \geq 2$. The following conditions are equivalent and if any is satisfied, then $\mathfrak{M}$ is said to be* Ivanov–Petrova.

1. *$\mathfrak{M}$ is spacelike Ivanov–Petrova.*

2. *$\mathfrak{M}$ is timelike Ivanov–Petrova.*

3. *$\mathfrak{M}$ is mixed Ivanov–Petrova.*

4. *For $2i \leq m$, there exist constants c_i so $\mathrm{Tr}\{\mathcal{A}(x, y)^{2i}\} = c_i\{\langle x, x \rangle \langle y, y \rangle - \langle x, y \rangle^2\}^i$ for all x, $y \in V$.*

Proof. Suppose that Assertions (1), (2), or (3) hold. Choose (p, q) appropriately; $\mathcal{O}_{(p,q)}$ is then a non-empty open subset of $V \oplus V$. By Lemma 1.11, there exist constants c_i so

$$\mathrm{Tr}\{\mathcal{A}(x, y)^{2i}\} = c_i\{\langle x, x \rangle \langle y, y \rangle - \langle x, y \rangle^2\}^i \text{ on } \mathcal{O}_{(p,q)}\ .$$

Lemma 1.12 then shows this equation holds on all of $V \oplus V$ which establishes Assertion (4). Conversely, if Assertion (4) holds, then $\mathrm{Tr}\{\mathcal{A}(x, y)^k\}$ is constant on $\mathcal{O}_{(p,q)}$ if k is even. By Lemma 1.5, $\mathrm{Tr}\{\mathcal{A}(x, y)^k\} = 0$ if k is odd. Consequently, Lemma 1.11 now implies Assertions (1), (2), and (3). $\square$

We have the following useful Corollary:

Corollary 1.23.   *Let $\mathfrak{M}$ be a curvature model of dimension m.*

1. *If $m = 3$, then $\mathfrak{M}$ is Ivanov–Petrova if and only if $\mathrm{Tr}\{\mathcal{A}(\pi)^2\}$ is constant on Gr_2^+.*

2. *If $m = 4$, then $\mathfrak{M}$ is Ivanov–Petrova if and only if $\left\{ \mathrm{Tr}\{\mathcal{A}(\pi)^2\}, \det\{\mathcal{A}(\pi)\} \right\}$ are constant on Gr_2^+.*

1.4.4   OSSERMAN IVANOV–PETROVA CURVATURE MODELS

The work in this section is also due to Esteban Calviño-Louzao [75] and treats curvature models of signature $(2, 2)$. Curvature models of constant sectional curvature are Ivanov–Petrova, but there are Ivanov–Petrova curvature models that do not have constant sectional curvature; Riemannian Ivanov–Petrova curvature models are completely classified in the Riemannian setting in dimensions $m \geq 4$ [137, 145, 159]. The situation is more complicated in neutral signature, as shown in Theorem 1.24 below. We recall the definitions of a para-Hermitian structure and of a hyper-para-Hermitian structure given in Section 1.3.4. Adopt the notation of Example 1.3.

Theorem 1.24.   *Let $\mathfrak{M} = (V, \langle \cdot, \cdot \rangle, A)$ be a curvature model of signature $(2, 2)$. Then $\mathfrak{M}$ is both Osserman and Ivanov–Petrova if and only if one of the following holds:*

1. *$A = \kappa A_{\langle \cdot, \cdot \rangle}$ for some constant κ.*

2. *There exists a pseudo-Hermitian structure J on $\mathfrak{M}$ so $A = \kappa (A_{\langle \cdot, \cdot \rangle} - \frac{1}{2} A_J)$ for some $\kappa \neq 0$.*

3. *There exists a para-Hermitian structure J on $\mathfrak{M}$ so $A = \kappa (A_{\langle \cdot, \cdot \rangle} + \frac{1}{2} A_J)$ for some $\kappa \neq 0$.*

4. *There exists a hyper-para-Hermitian structure $\{J_1, J_2, J_3\}$ on $\mathfrak{M}$ so $A = \kappa A_{J_1 + J_2}$ for some $\kappa \neq 0$.*

In this setting, the Jacobi operator is diagonalizable if and only if Case (4) does not pertain.

Curvature models of the form given by Case (2) above were previously reported in [172] in the Riemannian setting and in [264] for the signature $(2, 2)$ setting (see also [137]).

Remark 1.25.   In Cases (1)-(3), $\mathfrak{M}$ is Jordan Osserman and Jordan Ivanov–Petrova, while in Case (4) $\mathfrak{M}$ is Jordan Osserman and spacelike and timelike Jordan Ivanov–Petrova, but not mixed Jordan Ivanov-Petrova. If a curvature model $\mathfrak{M}$ is Jordan Ivanov-Petrova, the skew symmetric curvature operator has constant rank r on the set of non-degenerate 2-planes. If $\mathfrak{M}$ has constant sectional curvature, then $r = 2$. Otherwise, in Cases (2) and (3), $r = 4$. With respect to algebraic curvature tensors given by Case (4), they have 2-step nilpotent Jacobi operators and their skew symmetric curvature operators have constant rank 2 for oriented non-degenerate spacelike or timelike 2-planes, but rank changes from 0 to 2 for oriented non-degenerate mixed 2-planes.

We shall now analyze the Ivanov–Petrova condition for Osserman algebraic curvature tensors which have non-diagonalizable Jacobi operators. The following is a direct consequence of Theorem 1.24 and Remark 1.25:

Corollary 1.26.   *Let $\mathfrak{M}$ be a curvature model of signature $(2, 2)$. Then $\mathfrak{M}$ is Osserman and Ivanov–Petrova with non-diagonalizable Jacobi operator if and only if the Jacobi operator is a 2-step nilpotent operator.*

1.4.5   COMMUTING CURVATURE MODELS

Definition 1.27.   Let $\mathfrak{M} = (V, \langle \cdot, \cdot \rangle, A)$ be a curvature model. Then $\mathfrak{M}$ is said to be

1. A *curvature–curvature commuting* model if $\mathcal{A}(x, y)\mathcal{A}(z, w) = \mathcal{A}(z, w)\mathcal{A}(x, y) \ \forall x, y, z, w$.

2. A *curvature–Jacobi commuting* model if $\mathcal{J}(x)\mathcal{A}(y, z) = \mathcal{A}(y, z)\mathcal{J}(x) \ \forall x, y, z$.

3. A *curvature–Ricci commuting* model if $\mathcal{A}(x, y)\rho = \rho\mathcal{A}(x, y) \ \forall x, y$.

4. A *Jacobi–Jacobi commuting* model if $\mathcal{J}(x)\mathcal{J}(y) = \mathcal{J}(y)\mathcal{J}(x) \ \forall x, y$.

5. A *Jacobi–Ricci commuting* model if $\mathcal{J}(x)\rho = \rho\mathcal{J}(x) \ \forall x$.

Curvature–Ricci commuting curvature models are also known in the literature as *Ricci semi-symmetric curvature models* [1]. The conditions in Definition 1.27 have also been described elsewhere in the literature as "skew–Tsankov", as "mixed–Tsankov", as "skew–Videv", as "Jacobi–Tsankov", and as "Jacobi–Videv", respectively, and the general field of investigation of such conditions is often referred to as Stanilov–Tsankov–Videv theory [48]. We have chosen to change the notation from that employed previously to put these conditions in parallel as much as possible. One has the following result [150]:

Theorem 1.28.   *Let $\mathfrak{M}$ be a curvature model.*

1. *The following assertions are equivalent:*

   (a) $\mathfrak{M}$ is a Jacobi–Ricci commuting model.

   (b) $\mathfrak{M}$ is a curvature–Ricci commuting model.

   (c) $A(\rho x, y, z, w) = A(x, \rho y, z, w) = A(x, y, \rho z, w) = A(x, y, z, \rho w) \ \forall x, y, z, w$.

2. *The following assertions are equivalent:*

   (a) $\mathfrak{M}$ is a curvature–Jacobi commuting model.

   (b) $\mathfrak{M}$ is a Jacobi–Jacobi commuting model.

Clearly, any Einstein metric is curvature–Ricci commuting but there are many other non-trivial examples [50]. The notion of curvature-Ricci commuting is a generalization of the semi-symmetric condition (see [1] and the references therein). Semi-symmetric manifolds of conullity two are curvature–curvature commuting [38, 54]. Besides the progress made [150, 152], as far as we know a complete description of curvature–Ricci commuting manifolds is not yet available.

In the Lorentzian category, Jacobi–Jacobi commuting (or equivalently curvature–Jacobi commuting) curvature models are necessarily flat [53, 150]. However, the geometrical significance of the curvature–curvature commuting condition is not yet well-understood although some progresses have been made in the Riemannian setting [54].

Example 1.29.   Let $\mathfrak{M}_0 := (V_0, \langle \cdot, \cdot \rangle, A_0)$ be a Riemannian Einstein curvature model. Let $\{e_i\}$ be an orthonormal basis for V_0. Let $V_1 = V_0^+ \oplus V_0^-$ be two copies of V_0 with bases $\{e_i^+, e_i^-\}$. Let $\mathfrak{M}_1 := (V_1, \langle \cdot, \cdot \rangle, A_1)$ where

$$\langle e_i^+, e_i^+ \rangle = 1, \quad \langle e_i^-, e_i^- \rangle = -1,$$

$$A_1(e_i^-, e_j^+, e_k^+, e_\ell^+) = A_1(e_i^+, e_j^-, e_k^+, e_\ell^+) = A_1(e_i^+, e_j^+, e_k^-, e_\ell^+)$$
$$= A_1(e_i^+, e_j^+, e_k^+, e_\ell^-) = A_0(e_i, e_j, e_k, e_\ell),$$

$$A_1(e_i^+, e_j^-, e_k^-, e_\ell^-) = A_1(e_i^-, e_j^+, e_k^-, e_\ell^-) = A_1(e_i^-, e_j^-, e_k^+, e_\ell^-)$$
$$= A_1(e_i^-, e_j^-, e_k^-, e_\ell^+) = -A_0(e_i, e_j, e_k, e_\ell).$$

Then $\mathfrak{M}_1$ is Jacobi–Ricci commuting and ρ^2 is a negative multiple of Id. We refer to [150] for further details. Examples of this type can also be realized geometrically – see the discussion in Section 6.5.4; this condition is closely related to the decomposition factor $\mathcal{W}_7$ of Theorem 1.8.

CHAPTER 2

Basic Geometrical Notions

2.1 INTRODUCTION

In this Chapter, we turn our attention to geometric structures. We begin in Section 2.2 with a bit of history relating the study of pseudo-Riemannian manifolds to mathematical physics.

Section 2.3 is an introduction to the theory of smooth manifolds. We discuss the tangent bundle and the Lie bracket. We define the flow of a smooth vector field and the Lie derivative. The Lie algebra of a Lie group is presented and the unit sphere in the quaternions is given as an example. We treat the geometry of the cotangent bundle, canonical coordinates, and symplectic forms. Torsion free connections, the curvature operator, the Jacobi operator, and the Ricci tensor are introduced. The Weyl projective operator is defined and various geometric realizability theorems are presented. Geodesics and parallel transport are defined and as is the holonomy group.

Section 2.4 deals with pseudo-Riemannian geometry. The Levi-Civita connection, the Christoffel symbols of the first and second kind, the Weyl conformal curvature tensor, and theorems of geometric realizability are discussed. We make contact with notions of Section 1.4 and introduce associated operators in the geometric setting. Weyl scalar invariants and null distributions are presented as are results concerning pseudo-Riemannian holonomy.

Section 2.5 involves additional geometric structures. We discuss pseudo-Hermitian structures, para-Hermitian structures, and the Gray identity. We also present additional geometric realization theorems. Homogeneous and symmetric spaces, as well as curvature homogeneity are treated.

2.2 HISTORY

Riemannian, pseudo-Riemannian, and affine geometry are central areas of mathematics – see, for example the following basic references [26, 78, 182, 191, 196, 215, 231, 239]. Although Riemannian geometry is the most prevalent, in addition to their purely intrinsic interest, pseudo-Riemannian manifolds appear in mathematical physics. Lorentzian geometry is, of course, intimately linked with general relativity. There are, however, higher signature applications. They play a role in general relativity [47, 119], they are important in string theory [22, 82, 176, 195, 216], and they also are important in braneworld cosmology [230].

2.3 BASIC MANIFOLD THEORY

Let $x = (x_1, \ldots, x_m)$ be a system of local coordinates on a smooth manifold M of dimension m. The collection $\{\partial_{x_1}, \ldots, \partial_{x_m}\}$ is the coordinate frame for the tangent bundle TM where we set $\partial_{x_i} := \frac{\partial}{\partial x_i}$. Similarly, the collection $\{dx_1, \ldots, dx_m\}$ is the coordinate frame for the cotangent bundle T^*M. If

$h \in C^\infty(M)$ is a smooth function, let

$$h_{i_1 \ldots i_r} := \frac{\partial^r h}{\partial x_{i_1} \ldots \partial x_{i_r}} \ .$$

A *foliation* of dimension k on M is a collection of submanifolds $\mathcal{F}$ (*leaves*) which are disjoint, connected and immersed in M. Moreover, the union of the elements of $\mathcal{F}$ is M and for every $P \in M$ there is a chart whose intersection with each leaf is either the empty set or a countable union of k dimensional slices with $x_{k+1} = \kappa_{k+1}, \ldots, x_m = \kappa_m$, where $\kappa_{k+1}, \ldots, \kappa_m$ are constants.

We say that M is *simply connected* if M is connected and if the fundamental group of M is trivial, i.e., every closed curve in M is contractible to a point.

2.3.1   THE TANGENT BUNDLE, LIE BRACKET, AND LIE GROUPS

Let $X, Y \in C^\infty(TM)$ be smooth vector fields on M. The *Lie bracket* of X and Y is the vector field characterized by the identity:

$$([X, Y])f = (XY - YX)f \quad \forall f \in C^\infty(M) \ .$$

As mixed partial derivatives commute, $[\partial_{x_i}, \partial_{x_j}] = 0$. Expand $X = \sum_i a_i \partial_{x_i}$ and $Y = \sum_j b_j \partial_{x_j}$. We then have that:

$$[X, Y] = \sum_{i,j} \left\{ a_i \partial_{x_i} b_j - b_i \partial_{x_i} a_j \right\} \partial_{x_j} \ .$$

It is immediate from the definition that the Lie bracket $[\cdot, \cdot]$ is skew symmetric and that it satisfies the *Jacobi identity*:

$$[X, [Y, Z]] + [Y, [Z, X]] + [Z, [X, Y]] = 0 \ . \tag{2.1}$$

More generally, a vector space $\mathfrak{g}$ is said to be a *Lie algebra* if there is a bilinear map $[\cdot, \cdot]$ from $\mathfrak{g} \times \mathfrak{g}$ to $\mathfrak{g}$ such that $[X, Y] = -[Y, X]$ and so that Equation (2.1) is satisfied. It is then immediate that the Lie bracket gives $C^\infty(TM)$ the structure of a Lie algebra.

Let X be a smooth vector field on M and let C be a compact subset of M. The *flow* defined by X is a smooth map

$$\Phi : C \times [0, \epsilon] \to M$$

defined for some $\epsilon = \epsilon(C, X) > 0$ which is characterized by the properties:

$$\Phi(x, 0) = x, \quad \Phi(\Phi(x, t), s) = \Phi(x, t + s), \quad \partial_t \Phi(x, t) = X(\Phi(x, t)) \ .$$

Let ξ be a tensor field on M. We use the flow Φ to pull back ξ to define a new tensor field $\Phi^*(\xi)(t)$. The *Lie derivative* is then defined by

$$\mathcal{L}_X \xi := \lim_{t \to 0} \frac{\Phi^*(\xi)(t) - \xi}{t} \ .$$

If $X \neq 0$, choose coordinates locally so that $X = \partial_{x_1}$. If $Y = \sum_i a_i \partial_{x_i}$ and if $\omega = \sum_j b_j dx_j$,

$$\mathcal{L}_X Y = \sum_i X(a_i)\partial_{x_i} \quad \text{and} \quad \mathcal{L}_X \omega = \sum_j X(b_j)dx_j \ .$$

A *Lie group* G is a smooth manifold that is also a group with smooth group operations; that is, the multiplication and the inverse maps are both smooth. The identity element of G is denoted by e. If $a \in G$, define *left multiplication* L_a and *right multiplication* R_a, respectively, by setting:

$$L_a : g \rightarrow ag \quad \text{and} \quad R_a : g \rightarrow ga \ .$$

The maps L_a and R_a are diffeomorphisms of G for any $a \in G$. A vector field $X \in C^\infty(TG)$ is said to be *left invariant* (resp. *right invariant*) if $L_a^* X = X$ (resp. $R_a^* X = X$) for all $a \in G$. Let $\mathfrak{g}$ be the vector space of all left invariant vector fields on G. The map $X \rightarrow X(e)$ identifies $\mathfrak{g}$ with $T_e G$. Since the Lie bracket of two left invariant vector fields is again left invariant, $\mathfrak{g}$ inherits the structure of a Lie algebra; the essence of Lie theory is to relate geometric and topological properties of the group G to algebraic properties of $\mathfrak{g}$. Every Lie group G admits non-trivial smooth left invariant measures and right invariant measures; we say that G is *unimodular* if G admits non-trivial smooth bi-invariant measures.

The *adjoint map* ad is defined by $\text{ad}_X(Y) = [X, Y]$. The *Cartan-Killing form* is

$$\omega(X, Y) := \text{Tr}\{\text{ad}_X \text{ad}_Y\} \ .$$

An example is perhaps helpful at this stage. Let S^3 be the sphere of unit quaternions;

$$S^3 = \{x_0 + ix_1 + jx_2 + kx_3 : x_0^2 + x_1^2 + x_2^2 + x_3^2 = 1\} \ .$$

Quaternion multiplication gives this the structure of a group. Let $\mathfrak{g} := \text{Span}_{\mathbb{R}}\{i, j, k\}$ be the purely imaginary quaternions. If $\xi \in \mathfrak{g}$, let $X_\xi(x) := x \cdot \xi$. Since $x \perp X_\xi(x)$, this is a tangent vector to S^3. Since left and right quaternion multiplication commute, the vector field X_ξ is left invariant and the map $\xi \rightarrow X_\xi$ identifies $\mathfrak{g}$ with the Lie algebra of S^3. We compute that

$$[X_i, X_j] = -2X_k, \quad [X_j, X_k] = -2X_i, \quad [X_k, X_i] = -2X_j \ .$$

It is now an easy matter to see that

$$\omega(X_i, X_i) = \omega(X_j, X_j) = \omega(X_k, X_k) = -4$$

while $\omega(X_i, X_j) = \omega(X_i, X_k) = \omega(X_j, X_k) = 0$. Thus, the Cartan-Killing form is negative definite in this instance.

2.3.2   THE COTANGENT BUNDLE AND SYMPLECTIC GEOMETRY

Let M be a smooth manifold of even dimension $m = 2n$. Let T^*M be the cotangent bundle. Let $\Lambda^p(T^*M)$ be the bundle of alternating p forms and let $S^p(T^*M)$ be the bundle of symmetric p forms. The case $p = 2$ is of particular importance; we may decompose

$$T^*M \otimes T^*M = \Lambda^2(T^*M) \oplus S^2(T^*M) \ .$$

If $\xi, \eta \in T^*M$, we define the *wedge product* "$\wedge$" and the *symmetric product* "$\circ$" by setting

$$\xi \wedge \eta := \tfrac{1}{2}(\xi \otimes \eta - \eta \otimes \xi) \in \Lambda^2(T^*M),$$
$$\xi \circ \eta := \tfrac{1}{2}(\xi \otimes \eta + \eta \otimes \xi) \in S^2(T^*M). \tag{2.2}$$

Let $\Theta \in \Lambda^2(T^*M)$. We say that Θ gives a *symplectic structure* to M if $\Theta^n \neq 0$.

Let $(x_1, \ldots, x_m)$ be a system of local coordinates on M. If $\omega \in T^*M$, we may expand $\omega = x_{i'} dx_i$ where we sum over repeated indices; the $(x_{1'}, \ldots, x_{m'})$ are the dual fiber coordinates and the collection of $2m$ functions $(x_1, \ldots, x_m, x_{1'}, \ldots, x_{m'})$ are the *canonical coordinates* on T^*M. There is a canonical symplectic structure on the cotangent bundle:

Lemma 2.1.   *Let* $\Omega := \sum_i dx_i \wedge dx_{i'}$. *This 2 form is independent of the coordinate system* $(x_1, \ldots, x_m)$ *and is called the* canonical symplectic form.

Proof. Let $(y_1, \ldots, y_m)$ be another system of local coordinates. We may expand:

$$dy_i = \sum_j \partial_{x_j} y_i \cdot dx_j .$$

Let $\omega \in T^*M$. Expand $\omega = \sum_i y_{i'} dy_i = \sum_{i,j} y_{i'} \partial_{x_j} y_i \cdot dx_j$. This shows $x_{j'} = \sum_i y_{i'} \partial_{x_j} y_i$. Thus:

$$\sum_j dx_j \wedge dx_{j'} = \sum_{i,j,k} y_{i'} \partial_{x_j} \partial_{x_k} y_i \cdot dx_j \wedge dx_k + \sum_{i,j} \partial_{x_j} y_i \cdot dx_j \wedge dy_{i'} = \sum_i dy_i \wedge dy_{i'}.$$

$\square$

If Ω is a closed symplectic form on M, then there are local coordinates so

$$\Omega = dx_1 \wedge dy_1 + \cdots + dx_n \wedge dy_n .$$

Thus, the local geometry is canonical and a model can be taken to be $T^*(\mathbb{R}^n)$ where $2n = m$.

For each vector field X on M, define a function $\iota X : T^*M \to \mathbb{R}$ by

$$\iota X(P, \omega) = \omega(X_P) .$$

We may expand $X = X^i \partial_{x_i}$. We then have that this tautological function is given by

$$\iota X(x_i, x_{i'}) = \sum_i x_{i'} X^i .$$

Let $\tilde{Y}, \tilde{Z} \in C^\infty(T(T^*M))$ be smooth vector fields on T^*M. Then

$$\tilde{Y} = \tilde{Z} \quad \Leftrightarrow \quad \tilde{Y}(\iota X) = \tilde{Z}(\iota X) \quad \text{for all} \quad X \in C^\infty(TM) .$$

The *complete lift* X^C is characterized by the identity

$$X^C(\iota Z) = \iota[X, Z] \quad \text{for all} \quad Z \in C^\infty(TM) . \tag{2.3}$$

A $(0, s)$ tensor field on T^*M is characterized by its evaluation on complete lifts of vector fields on M because

$$T_{(P,\omega)}(T^*M) = \{X^C_{P,\omega} : X \in C^\infty(TM)\} .$$

Let $T \in C^\infty(\mathrm{End}(TM))$ be an endomorphism of the tangent bundle of M. We define a 1 form $\iota T \in C^\infty(T^*(T^*M))$ which is characterized by the identity

$$\iota T(X^C) = \iota(TX) .$$

2.3.3   CONNECTIONS, CURVATURE, GEODESICS, AND HOLONOMY

Let D be a *connection* on M; $D : C^\infty(TM) \to C^\infty(T^*M \otimes TM)$ is a linear first order partial differential operator which satisfies the Leibnitz rule

$$D(fX) = df \otimes X + f DX .$$

We let $D_X Y = (X, DY)$ where $(\cdot, \cdot)$ indicates the natural pairing $TM \otimes T^*M \otimes TM \to TM$ given by evaluation on the first factor. We say that D is a *torsion free connection* if the torsion tensor $\mathcal{T} = \mathcal{T}_D$ vanishes where

$$\mathcal{T}(X, Y) := D_X Y - D_Y X - [X, Y] .$$

The *curvature operator* $\mathcal{R} = \mathcal{R}_D$ is defined by the formula

$$\mathcal{R}(X, Y)Z := (D_X D_Y - D_Y D_X - D_{[X,Y]})Z .$$

One computes easily that if $f \in C^\infty(M)$:

$$\mathcal{R}(fX, Y)Z = \mathcal{R}(X, fY)Z = \mathcal{R}(X, Y)fZ = f\mathcal{R}(X, Y)Z .$$

Thus, $\mathcal{R}(X, Y)Z$ at a point P only depends on $X(P)$, $Y(P)$, and $Z(P)$; thus $\mathcal{R}$ is a tensor

$$\mathcal{R} \in T^*M \otimes T^*M \otimes \mathrm{End}(TM) .$$

We shall assume that D a is torsion free connection henceforth and let (M, D) denote the corresponding *affine manifold*. We then have the symmetries:

$$\mathcal{R}(X, Y) = -\mathcal{R}(Y, X), \tag{2.4}$$
$$\mathcal{R}(X, Y)Z + \mathcal{R}(Y, Z)X + \mathcal{R}(Z, X)Y = 0. \tag{2.5}$$

Equation (2.4) is often called a $\mathbb{Z}_2$ symmetry while Equation (2.5) is the *first Bianchi identity*.

It is convenient to work in a purely algebraic setting. Let V be a finite dimensional vector space. We say that $\mathcal{A} \in V^* \otimes V^* \otimes \mathrm{End}(V)$ is a *generalized curvature operator* if $\mathcal{A}$ has the symmetries of

Equations (2.4) and (2.5) above. We say that such an $\mathcal{A}$ is *geometrically realizable* if there exists a point P of an affine manifold (M, D) and if there exists an isomorphism $\phi : T_P M \rightarrow V$ so that $\phi^* \mathcal{A} = \mathcal{R}_P$. The following result shows that Equations (2.4) and (2.5) generate all the universal symmetries of the curvature operators of torsion free connections; we refer to [151] for further details:

Theorem 2.2. *Every generalized curvature operator is geometrically realizable.*

The associated *Ricci tensor* $\rho = \rho_D$ is defined by

$$\rho(X, Y) := \text{Tr}\{Z \rightarrow \mathcal{R}_D(Z, X)Y\} \ .$$

Note that in this setting, ρ_D need not be symmetric. The corresponding *Jacobi operator* $\mathcal{J} = \mathcal{J}_D$ is defined by

$$\mathcal{J}(X) : Z \rightarrow \mathcal{R}(Z, X)X \ .$$

Let $\mathfrak{F}(V) \subset V^* \otimes V^* \otimes \text{End}(V)$ be the vector space of all generalized curvature operators. The Ricci tensor defines a short exact sequence

$$0 \rightarrow \ker(\rho) \rightarrow \mathfrak{F}(V) \rightarrow V^* \otimes V^* \rightarrow 0 \ .$$

Let ρ_a and ρ_s be the components in $\Lambda^2(V^*)$ and $S^2(V^*)$, respectively, where

$$\rho_a(x, y) := \tfrac{1}{2}\{\rho(x, y) - \rho(y, x)\} \quad \text{and} \quad \rho_s(x, y) := \tfrac{1}{2}\{\rho(x, y) + \rho(y, x)\} \ .$$

The map ρ is equivariantly split by the map σ where

$$\{\sigma\rho_a\}(x, y)z = \tfrac{-1}{1+m}\{2\rho_a(x, y)z + \rho_a(x, z)y - \rho_a(y, z)x\},$$
$$\{\sigma\rho_s\}(x, y)z = \tfrac{1}{1-m}\{\rho_s(x, z)y - \rho_s(y, z)x\} \ .$$

We may then define the *Weyl projective curvature operator*

$$\mathcal{P}_{\mathcal{A}}(x, y)z := \mathcal{A}(x, y)z - \sigma\rho \in \ker \rho \ .$$

Theorem 2.3. *We have a decomposition of $\mathfrak{F}(V)$ into irreducible* $\text{GL}(V)$ *modules:*

$$\mathfrak{F}(V) = \ker(\rho) \oplus S^2(V^*) \oplus \Lambda^2(V^*) \ .$$

There are 8 natural geometric realization questions which arise in this context and whose realizability [154] may be summarized in the following table – the possibly non-zero components being indicated by $\star$:

$\ker(\rho)$	$S^2(V^*)$	$\Lambda^2(V^*)$		$\ker(\rho)$	$S^2(V^*)$	$\Lambda^2(V^*)$	
$\star$	$\star$	$\star$	yes	0	$\star$	$\star$	yes
$\star$	$\star$	0	yes	0	$\star$	0	yes
$\star$	0	$\star$	yes	0	0	$\star$	no
$\star$	0	0	yes	0	0	0	yes

Thus, for example, if the generalized curvature operator $\mathcal{A}$ is projectively flat (i.e., $\mathcal{P}_A = 0$) and Ricci symmetric (i.e., $\rho \in S^2(V^*)$), then $\mathcal{A}$ can be geometrically realized by a projectively flat Ricci symmetric torsion free connection . But if $\mathcal{A} \neq 0$ is projectively flat and Ricci antisymmetric, then $\mathcal{A}$ can not be geometrically realized by a projectively flat Ricci anti-symmetric torsion free connection.

Let (M, D) be an affine manifold. We say that a curve $\gamma(t)$ is an *affine geodesic* if the *geodesic equation* is satisfied:

$$D_{\dot{\gamma}} \dot{\gamma} = 0 \ .$$

Given a point $P \in M$, and an initial direction $X \in T_P M$, there is a unique curve γ with $\gamma(0) = P$ and $\dot{\gamma}(0) = X$ which satisfies the geodesic equation; γ need not be defined for all $t \in \mathbb{R}$ but γ does exist on a maximal interval $[0, T)$ for some $T > 0$. We say that (M, D) is *geodesically complete* if all geodesics exist for all time.

If $\gamma : [a, b] \to M$ is a smooth curve and if $X \in T_{\gamma(a)} M$ is given, then the *parallel transport* of X along γ is given by solving the following equation

$$\nabla_{\gamma(t)} X(t) = 0 \ .$$

Let $\gamma(a) = \gamma(b) = P$ be the basepoint of the manifold. The map $X(a) \to X(b)$ given by parallel translation around γ defines an invertible linear map $L_\gamma : T_P M \to T_P M$. The set of all such linear maps forms a group called the *holonomy group* of the connection and is denoted by $\mathcal{H}_D$; the holonomy groups corresponding to different points in a connected manifold are all isomorphic and thus the role of the basepoint P is usually suppressed. The holonomy group is a closed subgroup of $\mathrm{GL}(T_P M)$ and therefore is a Lie group.

2.4   PSEUDO-RIEMANNIAN GEOMETRY

Let $\mathcal{M} := (M, g)$ be a pseudo-Riemannian manifold. Here g is a symmetric non-degenerate smooth bilinear form on TM of signature (p, q) where $p + q = m$. We say that $\mathcal{M}$ is *Riemannian* if $p = 0$, i.e., if g is positive definite, that $\mathcal{M}$ is *Lorentzian* if $p = 1$, and that $\mathcal{M}$ has *neutral signature* if $p = q$.

Let "$\circ$" be the symmetric product defined in Equation (2.2). If $(x_1, \ldots, x_m)$ is a system of local coordinates on M, we may express

$$g = \sum_{i,j} g_{ij} dx_i \circ dx_j \quad \text{where} \quad g_{ij} := g(\partial_{x_i}, \partial_{x_j}) \ .$$

The pseudo-Riemannian element of volume is then given by

$$|\operatorname{dvol}| = \sqrt{|\det(g_{ij})|} dx_1 \ldots dx_m \ .$$

2.4.1   THE LEVI-CIVITA CONNECTION

The associated *Levi-Civita connection* $\nabla = \nabla^g$ is a connection on TM which is characterized by the following two properties:

(1) ∇ is torsion free, i.e., $\nabla_X Y - \nabla_Y X = [X, Y]$.
(2) ∇ is Riemannian, i.e., $g(\nabla X, Y) + g(X, \nabla Y) = dg(X, Y)$.

Suppose g is a Riemannian metric, i.e., g is positive definite. The *length of a curve γ* is given by:

$$L(\gamma) = \int \sqrt{g(\dot{\gamma}, \dot{\gamma})(t)} dt \ .$$

The distance between two points $P, Q \in M$ is then defined to be

$$d_g(P, Q) := \inf_{\gamma:\gamma(a)=P, \gamma(b)=Q} L(\gamma) \ .$$

One can show that this is in fact a true distance function; the topology defined by d_g and the original topology on M coincide. Geodesics locally minimize distance; conversely, if γ is a curve minimizing distance between points on γ, then up to a reparametrization, γ is a geodesic. In the pseudo-Riemannian setting where g is indefinite, there can be null curves and the connection between distance minimization and geodesy is lost.

We adopt the notation of Section 2.3.3. Let $\mathcal{M}$ be a pseudo-Riemannian manifold. We say that $\mathcal{M}$ exhibits *Ricci blowup* if there exists a geodesic γ defined for $t \in [0, T)$ with $T < \infty$ and if $\lim_{t \to T} |\rho(\dot{\gamma}, \dot{\gamma})| = \infty$ where ρ is the Ricci tensor. Clearly, if $\mathcal{M}$ exhibits Ricci blowup, then it is geodesically incomplete and it can not be isometrically embedded as an open subset of a geodesically complete manifold.

We expand

$$\nabla_{\partial_{x_i}} \partial_{x_j} = \Gamma_{ij}{}^k \partial_{x_k}$$

to define the *Christoffel symbols* of the first kind; ∇ is completely determined by these symbols. Let $g_{ij} := g(\partial_{x_i}, \partial_{x_j})$ and let g^{ij} be the inverse matrix. We use the metric to lower indices to define the Christoffel symbols of the second kind

$$\Gamma_{ijk} = g_{k\ell} \Gamma_{ij}{}^\ell = g(\nabla_{\partial_{x_i}} \partial_{x_j}, \partial_{x_k}) = \tfrac{1}{2} \{ \partial_{x_i} g_{jk} + \partial_{x_j} g_{ik} - \partial_{x_k} g_{ij} \} \ .$$

Let $\mathcal{R}$ be the curvature operator defined by ∇; we use the metric tensor to lower indices and define a corresponding curvature tensor $R \in \otimes^4(T_P^* M)$ by setting:

$$R(x, y, z, w) = g(\mathcal{R}(x, y)z, w) \ .$$

We then have that R satisfies the symmetries of Equation (1.2). Thus, we may define a curvature model by setting

$$\mathfrak{M}_P(\mathcal{M}) := (T_P M, g_P, R_P) \ .$$

Additionally, the Riemann curvature tensor also satisfies the *second Bianchi identity*,

$$(\nabla_X R)(Y, Z, U, T) + (\nabla_Y R)(Z, X, U, T) + (\nabla_Z R)(X, Y, U, T) = 0.$$

We say that a curvature model $\mathfrak{M} = (V, \langle \cdot, \cdot \rangle, A)$ is *geometrically realizable* if there is a point P of a pseudo-Riemannian manifold $\mathcal{M}$ and an isomorphism $\phi : T_P M \to V$ so that $\phi^*\langle \cdot, \cdot \rangle = g_P$ and $\phi^* A = R_P$. The following result shows every curvature model is geometrically realizable; in particular, the symmetries of Equation (1.2) generate the universal symmetries of the curvature tensor of the Levi-Civita connection.

Theorem 2.4. *Let $\mathfrak{M}$ be a curvature model. There exists a pseudo-Riemannian manifold $\mathcal{M}$, a point $P \in M$, and an isomorphism ϕ from $T_P M$ to V so that $\mathfrak{M}_P(\mathcal{M}) = \phi^*\mathfrak{M}$.*

Proof. Let $\{e_i\}$ be a basis for V. Let $A_{ijk\ell} := A(e_i, e_j, e_k, e_\ell)$ and set

$$g_{ik} := \langle e_i, e_k \rangle - \tfrac{1}{3} A_{ij\ell k} x^j x^\ell \ .$$

The symmetry $g_{ik} = g_{ki}$ is immediate. As $g_{ik}(0) = \langle e_i, e_k \rangle$, g is non-degenerate at the origin so g is a non-degenerate pseudo-Riemannian metric on some neighborhood of the origin.

$$\Gamma_{ijk} := g(\nabla_{\partial_{x_i}} \partial_{x_j}, \partial_{x_k}) = \tfrac{1}{2}(\partial_{x_i} g_{jk} + \partial_{x_j} g_{ik} - \partial_{x_k} g_{ij}) = O(|x|),$$

$$R_{ijk\ell}(0) = \{\partial_{x_i} \Gamma_{jk\ell} - \partial_{x_j} \Gamma_{ik\ell}\}(0)$$

$$= \tfrac{1}{2}\{\partial_{x_i} \partial_{x_k} g_{j\ell} + \partial_{x_j} \partial_{x_\ell} g_{ik} - \partial_{x_i} \partial_{x_\ell} g_{jk} - \partial_{x_j} \partial_{x_k} g_{i\ell}\}(0)$$

$$= \tfrac{1}{6}\{-A_{jik\ell} - A_{jki\ell} - A_{ij\ell k} - A_{i\ell jk} + A_{ji\ell k} + A_{j\ell ik} + A_{ijk\ell} + A_{ikj\ell}\}$$

$$= \tfrac{1}{6}\{4A_{ijk\ell} - 2A_{i\ell jk} - 2A_{ik\ell j}\} = A_{ijk\ell}. \qquad \square$$

2.4.2   ASSOCIATED NATURAL OPERATORS

The Jacobi operator, the skew symmetric curvature operator, the sectional curvature, the Ricci operator, the scalar curvature, the Weyl tensor, and the Schouten tensor extend to the geometric setting from the corresponding Equations (1.4), (1.5), (1.6), (1.7), (1.10), and (1.11).

We extend the properties discussed in Section 1.4 by using the curvature model $\mathfrak{M}_P(\mathcal{M})$. There is a slight fuss about local versus global properties. Thus, for example, we shall say that $\mathcal{M}$ is Osserman if the eigenvalues of the Jacobi operator are constant on the unit pseudo-sphere bundles $S^\pm(\mathcal{M})$ and that $\mathcal{M}$ is *pointwise Osserman* if the eigenvalues of the Jacobi operator are constant on the pseudo-spheres $S_P^\pm(\mathcal{M})$ for each point P in M but may vary, in principle, from point to point.

The Weyl conformal curvature tensor is a conformal invariant. One says (M, g) and $(M, \tilde{g})$ are *conformally equivalent* if $g = e^{2f} \tilde{g}$ for $f \in C^\infty(M)$. In this setting, we have

$$W_g(x, y, z, w) = e^{2f} W_{\tilde{g}}(x, y, z, w) \ .$$

In particular, (M, g) is conformally Osserman if and only if $(M, \tilde{g})$ is conformally Osserman. A manifold is conformally flat if and only if locally the metric tensor has the form

$$ds^2 = e^{2\phi}(-dx_1^2 - \cdots - dx_p^2 + dx_{p+1}^2 + \cdots + dx_m^2) \ .$$

The Schouten tensor defined in Equation (1.11) is Codazzi for every conformally flat pseudo-Riemannian manifold. Moreover, this is a sufficient condition for a 3 dimensional manifold to be conformally flat. For $m \geq 4$, an m dimensional manifold is locally conformally flat if and only if $W = 0$.

One has the following geometric realizability result [56]:

Theorem 2.5. *If $\mathfrak{M}$ is a conformally flat curvature model, then $\mathfrak{M}$ is geometrically realizable by a conformally flat manifold of constant scalar curvature.*

There is a natural operator that we shall be using throughout the book which has no algebraic analogue in Chapter 1. We define the *Szabó operator* as

$$\mathcal{S}(X)Y := (\nabla_X \mathcal{J})(X)Y = (\nabla_X \mathcal{R})(Y, X)X,$$

for any vector fields X and Y.

2.4.3   WEYL SCALAR INVARIANTS

Let $\nabla^k R_{i_1, \dots, i_4; j_1, \dots, j_k}$ be the components of the kth covariant derivative of the curvature operator defined by the Levi-Civita connection. Scalar invariants of the metric can be formed by using the metric tensors g^{ij} and g_{ij} to *fully contract* all indices. For example, the scalar curvature τ, the *norm of the Ricci tensor* $|\rho|^2$, and the *norm of the full curvature tensor* $|R|^2$ are scalar invariants which are given by:

$$\begin{aligned}
\tau &:= g^{ij} R_{kij}{}^k, \\
|\rho|^2 &:= g^{i_1 j_1} g^{i_2 j_2} R_{ki_1 i_2}{}^k R_{\ell j_1 j_2}{}^\ell, \quad \text{and} \\
|R|^2 &:= g^{i_1 j_1} g^{i_2 j_2} g^{i_3 j_3} g_{i_4 j_4} R_{i_1 i_2 i_3}{}^{i_4} R_{j_1 j_2 j_3}{}^{j_4} \ .
\end{aligned}$$

Such invariants are called *Weyl invariants*; if all possible such invariants vanish, then $\mathcal{M}$ is said to be *VSI* (vanishing scalar invariants). Let $\{x_1, \dots, x_n, y_1, \dots, y_n\}$ be coordinates on $\mathbb{R}^{2n}$. Let $\phi_{ij}(x) = \phi_{ji}(x)$ be a symmetric 2 tensor on $\mathbb{R}^n$. We adopt the notation of Section 3.5 and consider a *plane wave manifold* which is a twisted Riemannian extension of the form:

$$g := 2dx_i \circ dy_i + \phi_{ij}(x)dx_i \circ dx_j \ .$$

We refer to [148] for further details concerning plane wave manifolds. In Theorem 3.9, we will show that this manifold has vanishing scalar invariants and that it is not flat for generic ϕ as the (possibly) non-zero components of the curvature tensor are:

$$R(\partial_{x_i}, \partial_{x_j}, \partial_{x_k}, \partial_{x_\ell}) = \tfrac{1}{2}\{\partial_{x_i}\partial_{x_k}\phi_{j\ell} + \partial_{x_j}\partial_{x_\ell}\phi_{ik} - \partial_{x_i}\partial_{x_\ell}\phi_{jk} - \partial_{x_j}\partial_{x_k}\phi_{i\ell}\} \ .$$

Similarly, let (x_1, x_2, x_3) be the usual coordinates on $\mathbb{R}^3$. Let

$$g_f := 2dx_1 \circ dx_3 + dx_2 \circ dx_2 + f(x_2, x_3)dx_3 \circ dx_3 \ .$$

By Theorem 4.28, $(\mathbb{R}^3, g_f)$ is VSI; this manifold is not flat for generic f as

$$\mathcal{R}(\partial_{x_2}, \partial_{x_3})\partial_{x_2} = \tfrac{1}{2}f_{22}\partial_{x_1} \quad \text{and} \quad \mathcal{R}(\partial_{x_2}, \partial_{x_3})\partial_{x_3} = -\tfrac{1}{2}f_{22}\partial_{x_2} \ .$$

2.4.4 NULL DISTRIBUTIONS

Let $\mathcal{M}$ be a pseudo-Riemannian manifold of signature (p, q). We suppose given a splitting of the tangent bundle in the form $TM = V_1 \oplus V_2$ where V_1 and V_2 are smooth subbundles which are called *distributions*. This defines two complementary projections π_1 and π_2 of TM onto V_1 and V_2. We say that V_1 is a *parallel distribution* if $\nabla\pi_1 = 0$. Equivalently this means that if X_1 is any smooth vector field taking values in V_1, then ∇X_1 again takes values in V_1.

If $\mathcal{M}$ is Riemannian, we can take $V_2 = V_1^\perp$ to be the orthogonal complement of V_1 and in that case V_2 is again parallel. In the pseudo-Riemannian setting, of course, $V_2 \cap V_1$ need not be trivial and there exist examples where although V_1 is parallel, there exists no complementary parallel distribution as we shall see presently in Lemma 4.2.

Let V_1 be a parallel distribution. The rank of g restricted to V_1 is constant. We say that V_1 is a *null parallel distribution* if V_1 is parallel and if the metric restricted to V_1 vanishes identically.

Assume $\dim(V_1) = 1$ so that V_1 is a *line field*. Assume also that V_1 is parallel. If V_1 is not null at a single point, it is not null at any other point and parallel translation gives rise to a parallel vector field which spans V_1 if M is simply connected. However, if V_1 is null, it need not be spanned by a parallel vector field. Manifolds of dimension 3 which admit null parallel distributions are called 3 dimensional Walker manifolds; we return to this point in Section 4.3.

The complementary distribution of a parallel non-degenerate distribution is always integrable. This is not always the case if the parallel distribution $\mathcal{D}$ is degenerate. Indeed, the necessary and sufficient conditions for such integrability have been discussed in the literature (see, for example [20, 23, 90]). Let $\mathcal{M}$ be a pseudo-Riemannian manifold of neutral signature (n, n) admitting two parallel, complementary, and degenerate distributions $\mathcal{D}$ and $\mathcal{D}'$ (which implies $\dim\mathcal{D} = \dim\mathcal{D}' = n$). Then $\mathcal{R}(X, Y)Z \in \mathcal{D}$ (resp. $\mathcal{R}(X, Y)Z' \in \mathcal{D}'$) for all vector fields X, Y on M and any vector field $Z \in \mathcal{D}$ (resp. $Z' \in \mathcal{D}'$). Thus, both distributions remain invariant by the representation of the holonomy group G and therefore G becomes the connected subgroup of $SO_0(T_P M, g_P)$ generated by the matrices which leave $\mathcal{D}$ and $\mathcal{D}'$ invariant, i.e.,

$$G = \left\{ \begin{pmatrix} U & 0 \\ 0 & {}^tU^{-1} \end{pmatrix} : U \in GL^+(n, \mathbb{R}) \right\}.$$

Such a group G is the subgroup of $SO_0(n, n)$ of those elements which commute with the para-complex structure $J = \text{diag}[1, \dots, 1, -1, \dots, -1]$.

2.4.5   PSEUDO-RIEMANNIAN HOLONOMY

The holonomy group of an affine connection was discussed previously in Section 2.3.3. We use the Levi-Civita connection to define the holonomy group of a pseudo-Riemannian manifold. As the Levi-Civita connection is Riemannian, the parallel displacement preserves the metric. This implies that the holonomy group is a subgroup of $O(T_P M, g_P)$, and can be understood as a subgroup of $O(p, q)$ which is only defined up to conjugation in $O(p, q)$. This ensures that for a subspace $V \subset T_P M$ which is invariant under the holonomy group the orthogonal complement $V^\perp$ is invariant as well.

For a Riemannian metric the holonomy group acts completely reducibly, i.e., the tangent space decomposes into subspaces on which it acts trivially or irreducibly, but for indefinite metrics the situation is more subtle. We say that the holonomy group acts *indecomposably* if the metric is degenerate on any invariant proper subspace. In this case, we also say that the manifold is indecomposable. Of course, for Riemannian manifolds, indecomposability is equivalent to irreducibility.

A remarkable property is that the holonomy group of a product of Riemannian manifolds (i.e., equipped with the product metric) is the product of the holonomy groups of these manifolds (with the corresponding representation on the direct sum). Furthermore, a converse of this statement is true in the following sense. Let $\mathcal{M}$ be a connected pseudo-Riemannian manifold whose tangent space at a single point (and hence at every point) admits an orthogonal direct sum decomposition into non-degenerate subspaces which are invariant under the holonomy representation. Then $\mathcal{M}$ is locally isometric to a product of pseudo-Riemannian manifolds corresponding to the invariant subspaces. Moreover, the holonomy group is the product of the groups acting on the corresponding invariant subspaces. A global version of this statement was proven under the assumption that the manifold is simply connected and complete by G. de Rham [229] for Riemannian manifolds and by H. Wu [261] in arbitrary signature.

Theorem 2.6.   *Any simply connected complete pseudo-Riemannian manifold $\mathcal{M}$ is isometric to a product of simply connected complete pseudo-Riemannian manifolds one of which can be flat and the others have an indecomposably acting holonomy group. Moreover, the holonomy group of (M, g) is the product of these indecomposably acting holonomy groups.*

For indefinite metrics there exists the possibility that one of the factors in Theorem 2.6 is indecomposable, but not irreducible. This means that the holonomy representation admits an invariant subspace on which the metric is degenerate, but no proper non-degenerate invariant subspace. An attempt to classify holonomy groups for indefinite metrics has to provide a classification of these indecomposable, not irreducible holonomy groups. If a holonomy group acts indecomposably, but not irreducibly, with a degenerate invariant subspace $V \subset T_P M$, it admits a totally isotropic invariant subspace $I := V \cap V^\perp$ and thus $\mathcal{M}$ is a Walker manifold.

Walker coordinates, as we shall discuss presently in Theorem 3.1, have been extensively used in order to obtain metrics which realize the possible indecomposable, not irreducible holonomy groups.

2.5   OTHER GEOMETRIC STRUCTURES

In this section, we discuss additional geometric structures.

2.5.1   PSEUDO-HERMITIAN AND PARA-HERMITIAN STRUCTURES

Let $\mathcal{M}$ be a pseudo-Riemannian manifold. An *almost Hermitian* structure on $\mathcal{M}$ is an endomorphism J of the tangent bundle so that $J^*g = g$ and so $J^2 = -\operatorname{Id}$. Such a manifold necessarily is even dimensional and the signature satisfies $(p, q) = 2(\bar{p}, \bar{q})$ (see [163] for related work). There is an associated 2 form Ω, called the *almost Kaehler form*, which is defined (up to a possible sign convention) by

$$\Omega(x, y) := g(x, Jy) = g(Jx, J^2 y) = -g(Jx, y) = -\Omega(y, x) .$$

We can choose vector fields $\{e_1^-, \ldots, e_{\bar{p}}^-, e_1^+, \ldots, e_{\bar{q}}^+\}$ so that

$$\{e_1^-, Je_1^-, \ldots, e_{\bar{p}}^-, Je_{\bar{p}}^-, e_1^+, Je_1^+, \ldots, e_{\bar{q}}^+, Je_{\bar{q}}^+\}$$

is a local orthonormal frame for the tangent bundle. One then has that:

$$\Omega = -e_-^1 \wedge Je_-^1 - \cdots - e_-^{\bar{p}} \wedge Je_-^{\bar{p}} + e_+^1 \wedge Je_+^1 + \cdots + e_+^{\bar{q}} \wedge Je_+^{\bar{q}}$$

where

$$\{e_-^1, Je_-^1, \ldots, e_-^{\bar{p}}, Je_-^{\bar{p}}, e_+^1, Je_+^1, \ldots, e_+^{\bar{q}}, Je_+^{\bar{q}}\}$$

is the corresponding dual frame for the cotangent bundle. Thus, $\Omega^{\bar{p}+\bar{q}}$ is non-singular and defines a symplectic structure on M. One says that (M, g, J) is *almost Kaehler* if $d\Omega = 0$.

Let (M, g, J) be an almost Hermitian manifold of dimension $m = 2n$. We say that J is an *integrable Hermitian structure* or that (M, g, J) is a Hermitian manifold if there exist local coordinates $(x_1, y_1, \ldots, x_n, y_n)$ defined near every point of the manifold so that

$$J\partial_{x_i} = \partial_{y_i} \quad \text{and} \quad J\partial_{y_i} = -\partial_{x_i} .$$

The *Nijenhuis tensor* is defined for vector fields X and Y by

$$N_J(X, Y) := [X, Y] + J[JX, Y] + J[X, JY] - [JX, JY]. \tag{2.6}$$

Newlander and Nirenberg showed in [205] that the almost complex structure J is integrable if and only if $N_J = 0$. If $d\Omega = 0$ and if J is integrable then $\nabla J = 0$; in this setting the 2 form Ω is said to be Kaehler, the metric g is said to be Kaehler, and the manifold (M, g, J) is said to be a *Kaehler manifold*. If M is compact, this imposes significant topological restrictions on M. For example, the manifold $M = S^3 \times S^1$ admits an integrable complex structure and many Hermitian metrics but admits no Kaehler metric. A Hermitian manifold (M, g, J) is said to be *locally conformally Kaehler* if for any point $P \in M$ there exists an open neighborhood U and a function $f : U \to \mathbb{R}$ such that $(U, e^{2f}g, J)$ is a Kaehler manifold [109].

We use Equation (1.8) to define the $\star$-*scalar curvature* and the $\star$-*Ricci tensor*. We say that (M, g, J) is *weakly $\star$-Einstein* if $\rho^\star = \frac{\tau^\star}{m}g$. Schur's Lemma is not applicable in this context and $\tau^\star$ need not be constant. Thus, one says that (M, g, J) is $\star$-*Einstein* if additionally $\tau^\star$ is constant. Note that there exist weakly $\star$-Einstein manifolds which are not Einstein; see Theorem 5.18.

A pseudo-Riemannian metric g on an almost complex manifold (M, J) is said to be an *almost anti-Hermitian metric* if $J^*g = -g$. Additionally, (M, g, J) is said to be *anti-Kaehler* if the structure is parallel (i.e., $\nabla J = 0$) [45]. Associated to any almost anti-Hermitian structure (g, J) there is a metric $\phi(X, Y) = g(X, JY)$ which is usually referred to in the literature as the *twin metric*. The special significance of anti-Kaehler structures is that both g and ϕ share the same Levi-Civita connection without being (M, g) locally reducible.

An *almost para-Hermitian* structure on $\mathcal{M}$ is an endomorphism J of the tangent bundle so that $J^*g = -g$ and so $J^2 = \mathrm{Id}$; note that such a manifold $\mathcal{M}$ necessarily has neutral signature, i.e., $p = q = \frac{1}{2}m$. The notions $\star$-Einstein and weakly $\star$-Einstein are defined as in the almost Hermitian setting. As in the almost Hermitian setting, the associated *almost Kaehler form* Ω defines a symplectic structure; Ω is given by

$$\Omega(x, y) := g(x, Jy) = -g(Jx, J^2y) = -g(Jx, y) = -\Omega(y, x) \ .$$

(M, g, J) is said to be *almost para-Kaehler* if $d\Omega = 0$.

We say that J is an *integrable para-Hermitian structure* or is simply a *para-Hermitian structure* if the associated *Nijenhuis tensor*,

$$N_J(X, Y) := [X, Y] - J[JX, Y] - J[X, JY] + [JX, JY] \tag{2.7}$$

vanishes. In such a case (M, g, J) is said to be a *para-Hermitian manifold*. Analogously to the Hermitian setting, this condition is equivalent to the existence of local coordinate systems $(x_1, \ldots, x_n, y_1, \ldots, y_n)$ so that

$$J\partial_{x_i} = \partial_{y_i} \quad \text{and} \quad J\partial_{y_i} = \partial_{x_i} \ .$$

Finally, (M, g, J) is said to be *para-Kaehler* if J is integrable and $d\Omega = 0$ or, equivalently, $\nabla J = 0$. We refer to [120] for a discussion of almost para-Hermitian geometry from the point of view of general relativity.

2.5.2 HYPER-PARA-HERMITIAN STRUCTURES

An almost hyper-para-complex structure on a $4n$ dimensional manifold M is a triple

$$\mathcal{J} := \{J_1, J_2, J_3\}$$

where J_2, J_3 are almost para-complex structures (cf. [173]) and J_1 is an almost complex structure, satisfying the para-quaternionic identities

$$J_1^2 = -J_2^2 = -J_3^2 = -1, \qquad J_1 J_2 = -J_2 J_1 = J_3.$$

An almost hyper-para-Hermitian metric is a pseudo-Riemannian metric which is compatible with the almost hyper-para-complex structure, i.e.,

$$g(J_1 \cdot, J_1 \cdot) = -g(J_2 \cdot, J_2 \cdot) = -g(J_3 \cdot, J_3 \cdot) = g(\cdot, \cdot).$$

Such a structure is called hyper-para-Hermitian if all the structures J_i are integrable. If each J_i for $i = 1, 2, 3$, is parallel with respect to the Levi-Civita connection or, equivalently, the three Kaehler forms $\Omega_i(X, Y) = g(X, J_i Y)$ are closed, then the manifold is called hyper-symplectic [168] or hyper-para-Kaehler. In this case, J_2 and J_3 are para-Kaehler structures and it follows that g is a Walker manifold (see [177, 178] for more information).

If (g, J_1, J_2, J_3) is an almost hyper-para-Hermitian structure, on a manifold of dimension 4, then the bivectors corresponding via the metric to the 2 forms $\Omega_1, \Omega_2, \Omega_3$ define an orthonormal basis of Λ^-, and conversely, any orthonormal basis of Λ^- defines an almost hyper-para-Hermitian structure.

2.5.3   GEOMETRIC REALIZATIONS
We have the following geometric realization theorems [56]:

Theorem 2.7.

1. *Let $\mathfrak{M} = (V, \langle \cdot, \cdot \rangle, A)$ be a curvature model. Then $\mathfrak{M}$ is geometrically realizable by a pseudo-Riemannian manifold with constant scalar curvature.*

2. *Let $\mathfrak{C} = (V, \langle \cdot, \cdot \rangle, J, A)$ be a Hermitian curvature model. Then $\mathfrak{C}$ is geometrically realizable by an almost pseudo-Hermitian manifold with constant scalar curvature and constant $\star$-scalar curvature*

   .

3. *Let $\widetilde{\mathfrak{C}} = (V, \langle \cdot, \cdot \rangle, J, A)$ be a para-Hermitian curvature model. Then $\widetilde{\mathfrak{C}}$ is geometrically realizable by an almost para-Hermitian manifold with constant scalar curvature and constant $\star$-scalar curvature.*

This theorem generalizes appropriately to yield geometric realization theorems for hyper-Hermitian and hyper-para-Hermitian models. However, when we wish to impose the integrability condition to discuss realizations of Hermitian and para-Hermitian models, an additional condition arises which is called the *Gray condition* or the *Gray identity*. We refer to Gray [161] for the proof of the following result in the Hermitian setting and to [60] for the extension to the para-Hermitian setting:

Theorem 2.8.

1. *Let R be the curvature tensor of a Hermitian manifold* (M, g, J). *Then* $\forall x, y, z, w$

$$
\begin{aligned}
0 \;=\; & R(x, y, z, w) + R(Jx, Jy, Jz, Jw) - R(Jx, Jy, z, w) - R(Jx, y, Jz, w) \\
- \;& R(Jx, y, z, Jw) - R(x, Jy, Jz, w) - R(x, Jy, z, Jw) - R(x, y, Jz, Jw).
\end{aligned}
$$

2. *Let R be the curvature tensor of a para-Hermitian manifold* (M, g, J). *Then* $\forall x, y, z, w$

$$
\begin{aligned}
0 \;=\; & R(x, y, z, w) + R(Jx, Jy, Jz, Jw) + R(Jx, Jy, z, w) + R(Jx, y, Jz, w) \\
+ \;& R(Jx, y, z, Jw) + R(x, Jy, Jz, w) + R(x, Jy, z, Jw) + R(x, y, Jz, Jw).
\end{aligned}
$$

One has [57, 60] the following converse to Theorem 2.8:

Theorem 2.9.

1. *Let* $\mathfrak{C}$ *be a pseudo-Hermitian curvature model which satisfies the identity of* Theorem 2.8 (1). *Then* $\mathfrak{C}$ *is geometrically realizable by a pseudo-Hermitian manifold.*

2. *Let* $\tilde{\mathfrak{C}}$ *be a para-Hermitian curvature model which satisfies the identity of* Theorem 2.8 (2). *Then* $\tilde{\mathfrak{C}}$ *is geometrically realizable by a para-Hermitian manifold.*

We remark that A satisfies the Gray identity if and only if $A \perp \mathcal{W}_7$ where $\mathcal{W}_7$ is one of the factors in the Tricerri-Vanhecke curvature decomposition of Theorem 1.8. In Section 6.5.4 we will discuss a signature $(2, 2)$ locally symmetric manifold which has $\rho^2 = -\operatorname{Id}$ and has curvature tensor $A \in \mathcal{W}_7$.

2.5.4   HOMOGENEOUS SPACES, AND CURVATURE HOMOGENEITY

Let $\mathcal{M}$ be a pseudo-Riemannian manifold of signature (p, q). Let $\nabla^k R$ be the kth covariant derivative of the curvature tensor. The manifold $\mathcal{M}$ is said to be *locally symmetric* if $\nabla R = 0$. The manifold $\mathcal{M}$ is said to be *homogeneous* if the group of isometries acts transitively on $\mathcal{M}$; note that a simply connected complete locally symmetric space is homogeneous. $\mathcal{M}$ is said to be *locally homogeneous* provided that, for any points $P, Q \in M$, there exists an isometry ϕ mapping a neighborhood of P into a neighborhood of Q such that $\phi(P) = Q$.

One says that $\mathcal{M}$ is *curvature homogeneous* if given any two points P and Q of M, there is an isometry $\phi : T_P M \to T_Q M$ so that $\phi^* R_Q = R_P$, i.e., if the curvature looks the same at any two points of the manifold. Similarly, $\mathcal{M}$ is said to be *k curvature homogeneous* if in addition one has that $\phi^* \nabla^i R_Q = \nabla^i R_P$ for $1 \le i \le k$. Clearly, any locally homogeneous manifold is curvature homogeneous. What is perhaps somewhat surprising is that there are curvature homogeneous manifolds which are not locally homogeneous. We refer to [38, 64, 65, 149, 164, 235, 236, 262] for further details. One says that $\mathcal{M}$ is *k curvature modeled on a homogeneous space* $\mathcal{N}$ if for every point P

of M, there is an isometry $\phi : T_P M \to T_Q N$ so $\phi^* \nabla^i R_{N,Q} = \nabla^i R_{M,P}$ for $0 \le i \le k$; the point Q of N being irrelevant since $\mathcal{N}$ is assumed homogeneous. This implies $\mathcal{M}$ is k curvature homogeneous. We shall discuss curvature homogeneity further in Section 3.5 and in Section 4.7. We refer to [146, 147, 185, 202] for related work.

Remark 2.10.   In the Riemannian setting, Singer [240] showed there existed a universal bound k_m so that any Riemannian manifold which was k_m curvature homogeneous was locally homogeneous; this result was subsequently extended to the pseudo-Riemannian setting by Podesta and Spiro [224] and an analogous result established in the affine setting by Opozda [217]; see also [15].

Remark 2.11.   Prüfer, Tricerri, and Vanhecke [226] showed that if all local Weyl scalar invariants up to order $\frac{1}{2} m(m-1)$ are constant on a Riemannian manifold $\mathcal{M}$, then $\mathcal{M}$ is locally homogeneous and $\mathcal{M}$ is determined up to local isometry by these invariants. This fails in the pseudo-Riemannian setting. Recall, see the discussion in Section 2.4.3, that a manifold is said to be VSI if all the scalar Weyl invariants vanish. There are many non-flat manifolds which are VSI but which are not locally homogeneous as we shall see presently in Theorem 3.9. We also refer to the discussion [183, 225] for additional examples.

The following is a useful observation [21]

Lemma 2.12.   *Let $\mathcal{M}_i$ be real analytic pseudo-Riemannian manifolds for $i = 1, 2$. Assume there exist points $P_i \in M_i$ so $\exp_{M_i, P_i} : T_{P_i} M_i \to M_i$ is a diffeomorphism and so there exists an isomorphism Φ between $T_{P_1} M_1$ and $T_{P_2} M_2$ so $\Phi^* \nabla^k R_{P_2, M_2} = \nabla^k R_{P_1, M_1}$ for all k. Then we may define an isometry ϕ from $\mathcal{M}_1$ to $\mathcal{M}_2$ by setting:*

$$\phi := \exp_{\mathcal{M}_2, P_2} \circ \Phi \circ \exp_{\mathcal{M}_1, P_1}^{-1} \; .$$

2.5.5   TECHNICAL RESULTS IN DIFFERENTIAL EQUATIONS

We shall need the following results subsequently. We omit the proofs as they are elementary. Recall that $f_k := \partial_{x_k} f$.

Lemma 2.13.   *Let $\mathcal{O}$ be an open connected subset of $\mathbb{R}^4$. Let $p, q \in C^\infty(\mathcal{O})$ be functions only of (x_3, x_4). The following conditions are equivalent:*

1. *$p^2 = 2p_4$, $q^2 = 2q_3$, and $pq = p_3 + q_4$.*

2. *$p^2 = 2p_4$, $q^2 = 2q_3$, and $p_3 = q_4 = \frac{1}{2} pq$.*

3. *There exist $(a_0, a_3, a_4) \in \mathbb{R}^3 - \{0\}$ so that*

$$p = -2a_4(a_0 + a_3 x_3 + a_4 x_4)^{-1} \quad and \quad q = -2a_3(a_0 + a_3 x_3 + a_4 x_4)^{-1} \; .$$

Lemma 2.14.   *Let $\mathcal{O}$ be a connected open subset of $\mathbb{R}$ and let $h \in C^\infty(\mathcal{O})$. Assume that $h' \neq 0$ and that $hh''(h')^{-2}$ is constant. Then either $h(y) = ae^{by}$ or $h(y) = a(y+b)^c$.*

CHAPTER 3

Walker Structures

3.1 INTRODUCTION

In this Chapter, we outline the theory of Walker manifolds. In Section 3.2, we present a brief historical development and in Section 3.3, we treat Walker coordinates. Section 3.4 presents examples in very different natural geometric situations of strictly pseudo-Riemannian contexts where the underlying metric is a Walker one. More specifically we treat hypersurfaces with nilpotent shape operators, locally conformally flat metrics with nilpotent Ricci operator, degenerate pseudo-Riemannian homogeneous structures, para-Kaehler structures, and 2-step nilpotent Lie groups with degenerate center. The analysis of conformally symmetric metrics leads to the consideration of an important family of Walker manifolds: the Riemannian extensions.

A detailed description of Riemannian extensions is given in Section 3.5 (see also the results in Section 6.5). This construction, which relates affine and pseudo-Riemannian geometries, is very powerful in constructing new examples of strictly pseudo-Riemannian properties. We shall give a family of neutral signature Walker examples which are Riemannian extensions and relate them with curvature homogeneity. Given a torsion free connection D, a symmetric $(0, 2)$ tensor field ϕ on a manifold M, and $(1, 1)$ tensor fields T and S on M, there is a natural neutral signature metric $g_{D,\phi,T,S}$ on the cotangent bundle of M. We relate the Osserman and Ivanov–Petrova conditions for $g_{D,\phi}$ (a particular case of $g_{D,\phi,T,S}$) to conditions on the curvature of D. The case where D is flat gives rise to the notion of a plane wave manifold which we use to construct examples of nilpotent Walker manifolds. Osserman Riemannian extensions are investigated as is the relationship between a nilpotent Osserman manifold and an Osserman Riemannian extension. We establish interesting correspondences between torsion free connections with nilpotent skew symmetric curvature operator and Ivanov–Petrova Riemannian extensions.

3.2 HISTORICAL DEVELOPMENT

Walker manifolds constitute the underlying structure of many strictly pseudo-Riemannian situations with no Riemannian counterpart: indecomposable (but not irreducible) holonomy [23], Einstein hypersurfaces with nilpotent shape operators [194] or some classes of non-symmetric Osserman metrics [106] are typical examples. Walker manifolds have also been considered in general relativity in the study of $\mathfrak{hh}$ spaces [47, 119]. Moreover, the fact that para-Kaehler and hyper-symplectic metrics are necessarily of Walker type motivates the consideration of such metrics in connection with almost para-Hermitian structures.

Observe that there is a tight connection between Walker structures and Osserman metrics. First of all, it is a fact that most examples of Osserman metrics with non-diagonalizable Jacobi operator (i.e., Types II and III in Theorem 1.18) are realized as Walker manifolds. On the other hand, Walker manifolds appear as the underlying structure of several specific pseudo-Riemannian structures. We have already mentioned that the metric tensor of any para-Kaehler [92] (and hence any hyper-symplectic [168]) structure is necessarily of Walker type. The same occurs for the underlying metric of real hypersurfaces in indefinite space forms whose shape operator is nilpotent [192].

Lorentzian Walker manifolds have been studied extensively in the physics literature since they constitute the background metric of the pp-wave models [2, 179, 181, 200]. A pp-wave spacetime admits a covariantly constant null vector field U and therefore it is trivially recurrent (i.e., if one has $\nabla U = \omega \otimes U$ for some 1 form ω). Lorentzian Walker manifolds present many specific features both from the physical and geometric viewpoints [67, 80, 190, 225]. We also refer to related work [165, 166] for a discussion of generalized Lorentzian Walker manifolds (i.e., for spacetimes admitting a non-zero vector field n^a satisfying $R_{ijk\ell} n^\ell = 0$ or admitting a rank 2 symmetric or anti-symmetric tensor H with $\nabla H = 0$).

Riemannian extensions were originally defined by Patterson and Walker [221] and further investigated in [7] thus relating pseudo-Riemannian properties of T^*M with the affine structure of the base manifold (M, D). Moreover, Riemannian extensions were also considered in [100, 129] in relation to Osserman manifolds. The Riemannian extension provides a natural way of passing from the affine category in dimension n to the neutral signature pseudo-Riemannian category in dimension $m = 2n$.

3.3   WALKER COORDINATES

Walker [255] studied pseudo-Riemannian manifolds $\mathcal{M}$ with a parallel field of null planes $\mathcal{D}$ and derived a canonical form. Motivated by this seminal work, one says that a pseudo-Riemannian manifold $\mathcal{M}$ which admits a null parallel i.e., degenerate) distribution $\mathcal{D}$ is a *Walker manifold*.

Canonical forms were known previously for parallel non-degenerate distributions. In this case, the metric tensor, in matrix notation, expresses in canonical form as

$$(g_{ij}) = \begin{pmatrix} A & 0 \\ 0 & B \end{pmatrix},$$

where A is a symmetric $r \times r$ matrix whose coefficients are functions of $(x_1, \ldots, x_r)$ and B is a symmetric $(m - r) \times (m - r)$ matrix whose coefficients are functions of $(x_{r+1}, \ldots, x_m)$. Here m is the dimension of M and r is the dimension of the distribution $\mathcal{D}$.

Theorem 3.1.   *A canonical form for an m dimensional pseudo-Riemannian manifold $\mathcal{M}$ admitting a parallel field of null r dimensional planes $\mathcal{D}$ is given by the metric tensor in matrix form as*

$$(g_{ij}) = \begin{pmatrix} 0 & 0 & \mathrm{Id}_r \\ 0 & A & H \\ \mathrm{Id}_r & {}^tH & B \end{pmatrix},$$

where Id_r *is the* $r \times r$ *identity matrix and* A, B, H *are matrices whose coefficients are functions of the coordinates satisfying the following:*

1. A *and* B *are symmetric matrices of order* $(m - 2r) \times (m - 2r)$ *and* $r \times r$ *respectively.* H *is a matrix of order* $(m - 2r) \times r$ *and* $^t H$ *stands for the transpose of* H.

2. A *and* H *are independent of the coordinates* $(x_1, \ldots, x_r)$.

Furthermore, the null parallel r*-plane* $\mathcal{D}$ *is locally generated by the coordinate vector fields*

$$\{\partial_{x_1}, \ldots, \partial_{x_r}\} \ .$$

Proof. First observe that any pseudo-Riemannian manifold satisfying the conditions of the theorem admits a field of degenerate planes $\mathcal{D}$ spanned by $\{\partial_{x_1}, \ldots, \partial_{x_r}\}$. In order to show that $\mathcal{D}$ is parallel, we shall compute $\nabla_{\partial_{x_a}} \partial_{x_i}$ for all $i \in \{1, \ldots, r\}$, and $a \in \{1, \ldots, m\}$. Put

$$\nabla_{\partial_{x_a}} \partial_{x_i} = \Gamma_{ai}{}^j \partial_{x_j} + \Gamma_{ai}{}^\ell \partial_{x_\ell} + \Gamma_{ai}{}^\nu \partial_{x_\nu},$$

where $j \in \{1, \ldots, r\}, \ell \in \{r + 1, \ldots, m - r\}$ and $\nu \in \{m - r + 1, \ldots, m\}$. Now, a straightforward calculation using the fact that A and H are independent of the coordinates $(x_1, \ldots, x_r)$ shows that the Christoffel symbols $\Gamma_{ai}{}^\ell$ and $\Gamma_{ai}{}^\nu$ vanish identically. Hence, $\mathcal{D}$ is a parallel plane field.

Conversely, let $\mathcal{D}$ be an r dimensional parallel plane field. Then the orthogonal plane field $\mathcal{D}^\perp$ is also parallel and hence both $\mathcal{D}$ and $\mathcal{D}^\perp$ foliate M. As $\mathcal{D} \subset \mathcal{D}^\perp$ and $\dim \mathcal{D}^\perp = m - r$, there exist foliated coordinates

$$\overbrace{(x_1, \ldots, x_r, \underbrace{x_{r+1}, \ldots, x_{m-r}}, x_{m-r+1}, \ldots, x_m)}^{\mathcal{D}} \tag{3.1}$$

so that $\mathcal{D}$ (resp. $\mathcal{D}^\perp$) is locally spanned by $\{\partial_{x_1}, \ldots, \partial_{x_r}\}$ (resp. $\{\partial_{x_1}, \ldots, \partial_{x_{m-r}}\}$).

Now, since $\mathcal{D}$ is totally degenerate and $\mathcal{D}^\perp$ is the orthogonal complement of $\mathcal{D}$, one immediately has $g(\mathcal{D}, \mathcal{D}) = 0$ and $g(\mathcal{D}, \mathcal{D}^\perp) = 0$, which gives

$$g(\partial_{x_i}, \partial_{x_j}) = 0, \quad g(\partial_{x_i}, \partial_{x_\ell}) = g(\partial_{x_\ell}, \partial_{x_i}) = 0$$

for all $i, j \in \{1, \ldots, r\}, \ell \in \{r + 1, \ldots, m - r\}$.

Next consider the coordinate functions x_{m-r+i} and their gradients

$$E_i = \nabla x_{m-r+i},$$

$i \in \{1, \ldots, r\}$. Then one has

$$g(E_i, \partial_{x_\alpha}) = dx_{m-r+i}(\partial_{x_\alpha}) = \delta_\alpha^{m-r+i}$$

which shows that $g(E_i, \mathcal{D}^\perp) = 0$, and thus $E_i \in \mathcal{D}$ for all i. Since all the gradients are linearly independent, the parallel distribution is locally generated by the E_i's.

Now, it is shown in [255] (see also [20]) that $[E_i, E_j] = 0$ for all i, j, and thus the coordinates in Equation (3.1) can be further specialized so that $E_i = \partial_{x_i}$ for all $i \in \{1, \dots, r\}$. Then in the new system of coordinates, the matrix form of the metric becomes

$$(g_{ij}) = \begin{pmatrix} 0 & 0 & \mathrm{Id}_r \\ 0 & A & H \\ \mathrm{Id}_r & {}^t H & B \end{pmatrix}$$

where Id_r is the $r \times r$ identity matrix.

In what remains of the proof we will show that A and H are independent of the coordinates $(x_1, \dots, x_r)$. Let $i \in \{1, \dots, r\}$ and take $\alpha, \beta \in \{r+1, \dots, m-r\}$. Then

$$
\begin{aligned}
\partial_{x_i} g_{\alpha\beta} &= g(\nabla_{\partial_{x_i}} \partial_{x_\alpha}, \partial_{x_\beta}) + g(\partial_{x_\alpha}, \nabla_{\partial_{x_i}} \partial_{x_\beta}) \\
&= g(\nabla_{\partial_{x_\alpha}} \partial_{x_i}, \partial_{x_\beta}) + g(\partial_{x_\alpha}, \nabla_{\partial_{x_\beta}} \partial_{x_i}) = 0,
\end{aligned}
$$

since the distribution $\mathcal{D}$ is parallel (thus $g(\nabla \mathcal{D}, \mathcal{D}^\perp) = 0$). This shows that the matrix A does not depend on the coordinates $(x_1, \dots, x_r)$.

Similarly, for any $i \in \{1, \dots, r\}$, $\alpha \in \{r+1, \dots, m-r\}$ and $t \in \{m-r+1, \dots, m\}$, one has

$$
\begin{aligned}
\partial_{x_i} g_{\alpha t} &= g(\nabla_{\partial_{x_i}} \partial_{x_\alpha}, \partial_{x_t}) + g(\partial_{x_\alpha}, \nabla_{\partial_{x_i}} \partial_{x_t}) \\
&= g(\nabla_{\partial_{x_\alpha}} \partial_{x_i}, \partial_{x_t}) + g(\partial_{x_\alpha}, \nabla_{\partial_{x_t}} \partial_{x_i}) \\
&= -g(\partial_{x_i}, \nabla_{\partial_{x_\alpha}} \partial_{x_t}) \\
&= -g(\partial_{x_i}, \nabla_{\partial_{x_t}} \partial_{x_\alpha}) = 0,
\end{aligned}
$$

thus showing that the matrix H does not depend on the coordinates $(x_1, \dots, x_r)$. $\square$

Following the terminology of [255], a field of r-planes $\mathcal{D}$ is said to be *strictly parallel* if each vector in the plane at a point $P \in M$ is carried by parallel transport into a vector in the plane at another point $Q \in M$, the latter vector being the same for all paths from P to Q.

Theorem 3.2. *A canonical form for an m dimensional pseudo-Riemannian manifold M admitting a strictly parallel field of null r dimensional planes $\mathcal{D}$ is given by the metric tensor as in Theorem 3.1, where B is independent of the coordinates $(x_1, \dots, x_r)$.*

The local coordinates $(x_1, \dots, x_m)$ which give the canonical form of the metrics given in Theorem 3.1 or in Theorem 3.2 are not unique. Thus, care must be taken when defining invariants. Derdzinski and Roter gave a coordinate free version of Walker's theorem in [103] which avoids this difficulty. Their construction can be described as follows. Suppose that the following data are given:

1. Integers m and r with $0 < r \leq m/2$.

2. An r dimensional manifold Σ.

3. A bundle over Σ with some total space M, whose every fiber M_y, $y \in \Sigma$, is a $T_y^*\Sigma$-principal bundle over an $m - 2r$ dimensional manifold Q_y.

4. A pseudo-Riemannian metric h_y on each Q_y, $y \in \Sigma$.

All the y-dependent objects above, including the principal bundle structure, are assumed to vary smoothly with $y \in \Sigma$ and, in particular, the Q_y are the fibers of a bundle over Σ with a total space Q of dimension $m - r$.

Theorem 3.3.   *g and $\mathcal{D}$ are obtained by a natural construction (see [103] for details) from any prescribed data (m, r, Σ, M, Q, h), so that g is a pseudo-Riemannian metric on the m dimensional manifold U (where U is a non-empty open subset of M), and $\mathcal{D}$ is a g-null, g-parallel distribution of dimension r on U. Conversely, every null parallel distribution $\mathcal{D}$ on a pseudo-Riemannian manifold $\mathcal{M}$ is, locally and up to isometry, the result of applying the mentioned construction to some data (m, r, Σ, M, Q, h).*

Remark 3.4.   In order to clarify the data given above and relate it to the elements which appear in Walker's theorem, the following table associates the corresponding coordinates to each manifold. Assuming $(x_1, \ldots, x_m)$ are the coordinates for M in Walker's theorem, we have:

Manifold	Σ	Q	M_y	Q_y
Coordinates	$(x_{m-r+1}, \ldots, x_m)$	$(x_{r+1}, \ldots, x_m)$	$(x_1, \ldots, x_{m-r})$	$(x_{r+1}, \ldots, x_{m-r})$

Also note that the null parallel distribution $\mathcal{D}$ is spanned by $\{\partial_{x_1}, \ldots, \partial_{x_r}\}$.

Remark 3.5.   A different spinorial approach to Walker geometry has been developed in [189] (see also [85, 86]) which proved to be useful in describing 4 dimensional Walker manifolds of signature $(2, 2)$.

3.4   EXAMPLES OF WALKER MANIFOLDS

In this section, we shall present situations where Walker geometry arises naturally.

3.4.1   HYPERSURFACES WITH NILPOTENT SHAPE OPERATORS

We say $\mathcal{N}$ is a space form if $\mathcal{N} = \mathcal{N}(c)$ has constant sectional curvature c. Einstein hypersurfaces $\mathcal{M}$ in such a manifold $\mathcal{N}(c)$ have been studied by Magid [192], who showed that the shape operator S of any such hypersurface is diagonalizable, or it defines, after rescaling, a complex structure on M (i.e., $S^2 = -b^2$ Id for some $b \neq 0$), or it is a 2-step nilpotent operator (i.e., $S^2 = 0$, $S \neq 0$). The last two cases comprise the main differences with the theory of codimension one isometric immersions of Riemannian Einstein manifolds $\mathcal{M}$ into space forms of constant sectional curvature. Indeed, the case of diagonalizable shape operators has been already covered by the work of Fialkow [116]. Consequently, we shall pay special attention to the pseudo-Riemannian structure corresponding to the cases where the shape operator satisfies either $S^2 = -b^2$ Id for some $b \neq 0$, or $S^2 = 0$, $S \neq 0$.

It was shown by Magid [193] that Einstein hypersurfaces in indefinite space forms with imaginary principal curvatures have parallel second fundamental form, and thus the shape operator defines an anti-Kaehler structure on the hypersurface (cf. Section 2.5.1). Using this, he was able to show that such hypersurfaces are complex spheres $\mathbb{C}S^n(1/b)$ or $\mathbb{C}S^n(\sqrt{-1}/b)$, where

$$\mathbb{C}S^n(r) = \left\{ (z_1, \ldots, z_{n+1}) \in \mathbb{C}^{n+1} : z_1^2 + z_2^2 + \cdots + z_{n+1}^2 = r^2 \right\} \text{ for } r \in \mathbb{C} .$$

The later case ($S^2 = 0$, $S \neq 0$) is of main interest for the scope of this book. Since S is a self-adjoint operator, its kernel is a null i.e., a completely isotropic or a degenerate) subspace and it was shown in [194] that it is parallel whenever Rank S is maximal. This shows that the underlying metric on $\mathcal{M}$ is a Walker one. It is worth mentioning that a complete description of such hypersurfaces is not yet available.

An extension of the previous theory to the complex case was carried out in [204] where an analogous description of the shape operators has been obtained for complex Einstein hypersurfaces in indefinite complex space forms.

3.4.2   LOCALLY CONFORMALLY FLAT METRICS WITH NILPOTENT RICCI OPERATOR

By an abuse of notation, we shall also let ρ denote the *Ricci operator*; $g(\rho x, y) = \rho(x, y)$. Since the curvature operator $\mathcal{R}$ of a locally conformally flat manifold is essentially encoded by the Ricci operator, it is natural to search for classification results obtained by curvature conditions involving both $\mathcal{R}$ and ρ. One such condition is the commutativity of both the curvature and Ricci operators: $\mathcal{R}(X, Y) \cdot \rho = 0$, where X and Y run over all tangent vectors. Algebraic consequences of such condition have been obtained in [169] as follows.

Let $\mathcal{M}$ be a conformally flat pseudo-Riemannian manifold such that $\mathcal{R}(X, Y) \cdot \rho = 0$. Then the Ricci operator ρ satisfies one of the following conditions:

Type A: ρ is diagonalizable with respect to an orthonormal basis and has at most two real eigenvalues.
Type B: $\rho^2 = -a^2$ Id, where a is a non-zero real number.
Type C: $\rho^2 = 0$ and $\rho \neq 0$.

We continue our discussion of these three cases:

Type A. Only this case is available in the Riemannian setting. Manifolds of this type are necessarily one of the following:

1. A pseudo-Riemannian manifold of constant sectional curvature.

2. The product of two pseudo-Riemannian manifolds of non-zero constant sectional curvatures with opposite signs.

3. The product manifold of a pseudo-Riemannian manifold of non-zero constant sectional curvature and a 1 dimensional Lorentzian or Riemannian manifold.

Type B. More generally suppose $J^2 = -\operatorname{Id}$ and $J^*g = \pm g$. This corresponds to the following example. Give $\mathbb{C}^{n+1}$ the natural pseudo-Hermitian metric of neutral signature $(n+1, n+1)$. Let $\mathcal{M}$ be the complex hypersphere of dimension $m = 2n$ defined by the equation

$$z_1^2 + \cdots + z_{n+1}^2 = \sqrt{-1}b$$

where b is a non-zero real number. This manifold has the structure of the pseudo-Riemannian symmetric space $SO(n+1, \mathbb{C})/SO(n, \mathbb{C})$ and is diffeomorphic to the tangent bundle of the n dimensional sphere.

Type C. This corresponds to the case in which the manifold, on the open set where ρ has maximal rank, has a totally geodesic foliation with flat and complete leaves defined as the integral submanifolds of the involutive distribution given by the kernel of ρ. Moreover, if the maximal rank of ρ is 1 (which occurs for instance if the metric is Lorentzian), then both the kernel and the image of ρ are parallel degenerate distributions, which shows that the underlying metric is a Walker one.

3.4.3 DEGENERATE PSEUDO-RIEMANNIAN HOMOGENEOUS STRUCTURES

Ambrose and Singer [11] gave a characterization of simply connected and complete Riemannian homogeneous manifolds in terms of a $(1, 2)$ tensor field T on the manifold such that the connection $\bar{\nabla} = \nabla - T$ satisfies

$$\bar{\nabla}g = 0, \qquad \bar{\nabla}R = 0, \qquad \bar{\nabla}T = 0,$$

where ∇ is the Levi-Civita connection of the manifold. A tensor field T as above is usually called a *homogeneous structure*. Such characterization, later extended to reductive pseudo-Riemannian homogeneous spaces [122], was investigated in detail [123, 252]. A classification of such structures arises by considering the space of all covariant tensors $T_{XYZ} = g(T(X, Y), Z)$ which are skew symmetric in the last two arguments (which is equivalent to $\bar{\nabla}g = 0$),

$$\mathcal{T}(T_P M) = \{S \in \otimes^3 T_P^* M : S_{XYZ} = -S_{XZY}\}.$$

The action of the orthogonal group decomposes $\mathcal{T}(T_P M)$ into three invariant and irreducible components: $\mathcal{T}(T_P M) = \mathcal{T}_1 \oplus \mathcal{T}_2 \oplus \mathcal{T}_3$, where

$$\mathcal{T}_1 = \{T \in \mathcal{T}(T_P M) : T_{XYZ} = g(X,Y)\omega(Z) - g(X,Z)\omega(Y), \text{ for some } \omega \in T_P^* M\},$$

$$\mathcal{T}_2 = \{T \in \mathcal{T}(T_P M) : T_{XYZ} + T_{YZX} + T_{ZXY} = 0, \text{ and } \sum \varepsilon_i T_{e_i e_i Z} = 0\},$$

$$\mathcal{T}_3 = \{T \in \mathcal{T}(T_P M) : T_{XYZ} + T_{YXZ} = 0\}.$$

Here $\{e_i\}$ is an orthonormal basis and $\varepsilon_i = g(e_i, e_i)$.

Riemannian $\mathcal{T}_1$-homogeneous structures are very restrictive and indeed those which are not locally symmetric (i.e., $T \neq 0$) are necessarily of constant sectional curvature. This result has been extended to the pseudo-Riemannian case in [123] as follows:

Theorem 3.6. *Let $\mathcal{M} := (M, g)$ be a connected pseudo-Riemannian manifold which admits a non-degenerate homogeneous structure of type $\mathcal{T}_1$ defined by a vector field ξ i.e.,*

$$T(X,Y) = g(X,Y)\xi - g(\xi,Y)X \text{ where } g(\xi,\xi) \neq 0 .$$

Then $\mathcal{M}$ has constant sectional curvature equal to $-g(\xi,\xi)$.

The consideration of the degenerate case when ξ is a null vector has been settled [203]; the condition $\bar{\nabla} g = 0$ implies that $\nabla \xi = \alpha \otimes \xi$ for some 1 form α. This involved showing that ξ is a recurrent vector field. Consequently, the underlying metric is of Walker type. Moreover, just after renumbering the coordinates in [203], one has the following description of Walker coordinates.

Theorem 3.7. *An $n + 2$ dimensional pseudo-Riemannian manifold admits a degenerate homogeneous structure of type $\mathcal{T}_1$ if and only if there exist local coordinates $(u, x_1, \ldots, x_n, v)$ where the metric expresses as*

$$g = \begin{pmatrix} 0 & & 1 \\ & A & \\ 1 & & 2u + f \end{pmatrix}$$

where A is an $n \times n$ symmetric non-degenerate metric whose entries are functions of the coordinates $(x_1, \ldots, x_n)$, and f is a real valued function of the coordinates $(x_1, \ldots, x_n)$. Moreover, in such a case the vector field associated to the homogeneous structure is $\xi = \partial_u$.

3.4.4   PARA-KAEHLER GEOMETRY

Recall that a para-Kaehler manifold is a symplectic manifold admitting two transversal Lagrangian foliations (see [92, 173]). Such a structure induces a decomposition of the tangent bundle TM into the Whitney sum of Lagrangian subbundles L and L', that is, $TM = L \oplus L'$. By generalizing this definition, one sees that an almost para-Hermitian manifold may be defined to be an almost symplectic manifold (M, Ω) whose tangent bundle splits into the Whitney sum of Lagrangian subbundles. This definition implies that the $(1, 1)$ tensor field J defined by $J := \pi_L - \pi_{L'}$ is an almost para-complex structure, that is $J^2 = \mathrm{Id}$ on M, such that

$$\Omega(JX, JY) = -\Omega(X, Y) \quad \text{for all} \quad X, Y \in C^\infty(TM) \ .$$

Here π_L and $\pi_{L'}$ are the projections of TM onto L and L', respectively. The 2 form Ω induces a non-degenerate $(0, 2)$ tensor field g on M defined by

$$g(X, Y) := \Omega(X, JY) \ .$$

Now the relation between the almost para-complex and the almost symplectic structures on M shows that g defines a pseudo-Riemannian metric of signature (n, n) on M and moreover,

$$g(JX, Y) + g(X, JY) = 0 \quad \text{for all} \quad X, Y \in C^\infty(TM) \ .$$

The special significance of the para-Kaehler condition is equivalently stated in terms of the parallelizability of the para-complex structure with respect to the Levi-Civita connection of g, that is $\nabla J = 0$. The para-complex structure J has eigenvalues ± 1 with null i.e., completely isotropic) corresponding eigenspaces due to the skew symmetry of J. Moreover, since J is parallel in the para-Kaehler setting, the ± 1 eigenspaces are parallel as well. This shows that any para-Kaehler structure (g, J) arises from an underlying Walker manifold.

Para-Kaehler structures play an important role in pseudo-Riemannian holonomy since they correspond to the situation when the holonomy group leaves invariant two complementary totally degenerate subspaces (which correspond to the ± 1 eigenspaces of the para-complex structure J [23]). Hence, the complementary distribution of a Walker manifold is integrable if and only if the underlying structure is para-Kaehler. A canonical form of para-Kaehler metrics was given by Bérard Bergery and Ikemakhen, who showed the existence of coordinates $(x_1, \ldots, x_n, x_{1'}, \ldots, x_{n'})$ where the metric takes the form

$$(g_{ij}) = \begin{pmatrix} 0 & A \\ {}^t A & 0 \end{pmatrix} .$$

Here A is a symmetric matrix function with entries $a_{ij'} = \partial_{x_i} \partial_{x_j} h$ for some potential function h whose Taylor expansion starts with $x_1 x_{1'} + \cdots + x_n x_{n'}$ [23].

Note that the coordinates above are not Walker coordinates. We refer to Section 3.5.2 for a description of para-complex space forms in terms of Walker coordinates by means of modified Riemannian extensions of flat connections.

3.4.5   TWO-STEP NILPOTENT LIE GROUPS WITH DEGENERATE CENTER

Let N be a 2-step nilpotent Lie group with left invariant pseudo-Riemannian metric tensor $\langle \cdot, \cdot \rangle$ and Lie algebra $\mathfrak{n}$. In the Riemannian case, one splits $\mathfrak{n} = \mathfrak{z} \oplus \mathfrak{z}^\perp$ where the superscript denotes the orthogonal complement with respect to the inner product and $\mathfrak{z}$ stands for the center of $\mathfrak{n}$. In the pseudo-Riemannian case, however, $\mathfrak{z}$ may contain a degenerate subspace $\mathfrak{U}$ for which $\mathfrak{U} \subseteq \mathfrak{U}^\perp$. Hence, the following decomposition is introduced in [89]

$$\mathfrak{n} = \mathfrak{z} \oplus \mathfrak{b} = \mathfrak{U} \oplus \mathfrak{Z} \oplus \mathfrak{D} \oplus \mathfrak{E}$$

in which $\mathfrak{z} = \mathfrak{U} \oplus \mathfrak{Z}$ and $\mathfrak{b} = \mathfrak{D} \oplus \mathfrak{E}$. Here $\mathfrak{U}$ and $\mathfrak{D}$ are complementary null subspaces and one has that $\mathfrak{U}^\perp \cap \mathfrak{D}^\perp = \mathfrak{Z} \oplus \mathfrak{E}$. (Indeed, $\mathfrak{Z}$ is the part of the center in $\mathfrak{U}^\perp \cap \mathfrak{D}^\perp$ and $\mathfrak{E}$ is its orthogonal complement in $\mathfrak{U}^\perp \cap \mathfrak{D}^\perp$). The geometry of a pseudo-Riemannian 2-step nilpotent Lie group is essentially controlled by the linear mapping $j : \mathfrak{U} \oplus \mathfrak{Z} \to \operatorname{End}(\mathfrak{D} \oplus \mathfrak{E})$ defined by

$$\langle j(a)x, y \rangle = \langle [x, y], ia \rangle \ ,$$

where i is an involution interchanging $\mathfrak{U}$ and $\mathfrak{D}$. Now, since $[\mathfrak{n}, \mathfrak{n}] \subseteq \mathfrak{z}$, it immediately follows that $\mathfrak{U}$ is a parallel degenerate subspace and thus the metric $\langle \cdot, \cdot \rangle$ is necessarily a Walker one.

On the other hand, note that 4 dimensional indefinite Kaehler Lie algebras $\mathfrak{g}$ naturally split into two classes depending on whether a naturally defined Lagrangian ideal $\mathfrak{h}$ satisfying $\mathfrak{h} \cap J\mathfrak{h}$ is trivial or $\mathfrak{h} \cap J\mathfrak{h}$ coincides with $\mathfrak{g}$. If the second possibility occurs, then the induced metric is a Walker one. Such Lie algebras correspond to the cases $\mathbb{R} \times \mathfrak{h}_3$, $\mathfrak{aff}(\mathbb{C})$, $\mathfrak{r}_{4,-1,-1}$, $\delta_{4,1}$ and $\delta_{4,2}$. See [220] for details.

3.4.6   CONFORMALLY SYMMETRIC PSEUDO-RIEMANNIAN METRICS

A pseudo-Riemannian manifold $\mathcal{M}$ of dimension $m \geq 4$ is said to be *conformally symmetric* [81] if the Weyl tensor W of $\mathcal{M}$ is parallel. Obvious examples arise when $\mathcal{M}$ is conformally flat or locally symmetric. Conformally symmetric manifolds which are neither conformally flat nor locally symmetric have usually been referred to as *essentially conformally symmetric (ECS)*. Such manifolds must be non-Riemannian [102].

The local description of ECS manifolds naturally splits into two cases, depending on the dimension of a null parallel distribution. The *Olszak distribution*, which was introduced in [214] (see also [213]), is the subbundle $\mathcal{D}$ of TM such that the sections of $\mathcal{D}$ are precisely those vector fields U which satisfy the condition $g(U, \cdot) \wedge W(L, L', \cdot, \cdot) = 0$, i.e.,

$$\{g(U, X)W(L, L', Y, Z) + g(U, Y)W(L, L', Z, X) + g(U, Z)W(L, L', X, Y)\}X \wedge Y \wedge Z = 0$$

for all vector fields L, L', X, Y, Z. The distribution $\mathcal{D}$ is obviously parallel and its dimension is either 1 or 2 for any ECS manifold [104]. One may proceed as follows to describe the local structure of essentially conformally symmetric (ECS) manifolds.

ECS manifolds with dim $\mathcal{D} = 1$ are locally described by the following construction [104]. Let I be an open interval, let $f : I \to \mathbb{R}$ be a smooth function, and let V be a real vector space of dim $V = m - 2$ which is equipped with a pseudo-Euclidean inner product $\langle \cdot, \cdot \rangle$ and a non-zero trace free self-adjoint operator $A : (V, \langle \cdot, \cdot \rangle) \to (V, \langle \cdot, \cdot \rangle)$. Next define a pseudo-Riemannian manifold

$$(I \times \mathbb{R} \times V, \kappa dt^2 + dt ds + \gamma)$$

where (t, s) are the Cartesian coordinates in $I \times \mathbb{R}$, γ stands for the pullback of $\langle \cdot, \cdot \rangle$ to the manifold $I \times \mathbb{R} \times V$ and κ is the function given by

$$\kappa(t, s, \chi) = f(t)\langle \chi, \chi \rangle + \langle A\chi, \chi \rangle \quad \text{for any} \quad (t, s, \chi) \in I \times \mathbb{R} \times V .$$

The ECS manifolds with $\dim \mathcal{D} = 2$ are locally described by the following construction [104]. Let (Σ, D) be a projectively flat surface with a D-parallel volume form α and let $(V, \langle \cdot, \cdot \rangle)$ be an $m - 4$ dimensional vector space with a pseudo-Euclidean inner product. Further, let T be a twice-contravariant symmetric tensor field on Σ with $\operatorname{div}_D(\operatorname{div}_D T) + (\rho_D, T) = \pm 1$, where div_D denotes the D-divergence, ρ_D is the Ricci tensor of D and $(\cdot, \cdot)$ is the natural pairing. For such a T (which always exists for simply connected and non-compact Σ), define a twice-covariant symmetric tensor field τ on Σ by setting

$$\tau_{jk} = \alpha_{j\ell}\alpha_{ki}T^{\ell i} .$$

Let γ be the pseudo-Riemannian metric on V corresponding to $\langle \cdot, \cdot \rangle$ and let $\theta(v) = \langle v, v \rangle$ for all $v \in V$. Let g_D be the Riemannian extension of the torsion free connection D to $T^*\Sigma$. Consider

$$(T^*\Sigma \times V, g_D - 2\tau + \gamma - \theta\rho_D) .$$

We observe that if $\mathcal{D}$ has dimension 1 or 2, then the Olszak distribution $\mathcal{D}$ is null and parallel; thus the underlying metric is a Walker one.

3.5   RIEMANNIAN EXTENSIONS

In this section, we present basic information about Riemannian extensions and relate them to Walker geometry, to Ivanov–Petrova geometry, and to Osserman geometry. To be consistent with the classical notation for Riemannian extensions, as defined by Patterson and Walker [221] (see also [263]), throughout this section we shall renumber the Walker coordinates to fit the general form

$$(g_{ij}) = \begin{pmatrix} B & H & \operatorname{Id}_r \\ {}^tH & A & 0 \\ \operatorname{Id}_r & 0 & 0 \end{pmatrix} .$$

In the rest of this monograph we adopt the notation of Walker manifolds as in Section 3.3, so we will use coordinates $(x_1, \ldots, x_{2n})$ such that, after renumbering the indices appropriately, a Riemannian extension fit the general form

$$(g_{ij}) = \begin{pmatrix} 0 & \operatorname{Id}_r \\ \operatorname{Id}_r & \star \end{pmatrix} .$$

3.5.1   THE AFFINE CATEGORY

Let M be a smooth manifold of dimension n. We recall the discussion of the cotangent bundle given in Section 2.3.2. If $X \in C^\infty(TM)$ and $T \in C^\infty(\operatorname{End}(TM))$, recall that ιX, X^C, and ιT are characterized by the identities:

$$\iota X(P, \omega) = \omega(X_P), \quad X^C(\iota Z) = \iota[X, Z], \quad \iota T(X^C) = \iota(TX) .$$

Let D be a torsion free affine connection on M. The *Riemannian extension* is the pseudo-Riemannian metric g_D on T^*M of neutral signature (n, n) characterized by the identity:

$$g_D(X^C, Y^C) = -\iota(D_X Y + D_Y X) . \qquad (3.2)$$

Let $D_{\partial_{x_i}} \partial_{x_j} = {}^D\Gamma_{ij}{}^k \partial_{x_k}$ give the Christoffel symbols of the connection D. Then:

$$g_D = 2dx_i \circ dx_{i'} - 2x_{k'} {}^D\Gamma_{ij}{}^k dx_i \circ dx_j .$$

We can generalize this construction slightly. Let π be the canonical projection of the cotangent bundle T^*M on M. Let $\phi \in C^\infty(S^2(T^*M))$ be a symmetric $(0, 2)$ tensor field on M and let T and S belong to $C^\infty(\text{End}(TM))$. The *modified Riemannian extension* is the neutral signature metric on T^*M defined by

$$g_{D,\phi,T,S} := \iota T \circ \iota S + g_D + \pi^*\phi . \qquad (3.3)$$

In a system of local coordinates one has

$$g_{D,\phi,T,S} = 2dx_i \circ dx_{i'} + \{\tfrac{1}{2}x_{r'}x_{s'}(T_i^r S_j^s + T_j^r S_i^s) + \phi_{ij}(x) - 2x_{k'} {}^D\Gamma_{ij}{}^k\}dx_i \circ dx_j .$$

We set $T = S = 0$ to define the *twisted Riemannian extension*,

$$g_{D,\phi} := g_D + \pi^*\phi . \qquad (3.4)$$

Let $\mathcal{R}$ be the curvature Let $\tilde{\mathcal{R}}$ be the curvature operator of the Levi-Civita connection defined by $g_{D,\phi}$ and let $\mathcal{R}$ be the curvature of D. Express

$$\mathcal{R}(\partial_{x_i}, \partial_{x_j})\partial_{x_k} = \mathcal{R}_{ijk}{}^\ell \partial_{x_\ell} .$$

We refer to [263] for the following result:

Lemma 3.8. *The (possibly) non-zero curvatures of $\tilde{\mathcal{R}}$ are:*

$$\tilde{\mathcal{R}}_{kji}{}^h = \mathcal{R}_{kji}{}^h, \qquad \tilde{\mathcal{R}}_{kji}{}^{h'}, \qquad \tilde{\mathcal{R}}_{kji'}{}^{h'} = -\mathcal{R}_{kjh}{}^i, \qquad \tilde{\mathcal{R}}_{k'ji}{}^{h'} = \mathcal{R}_{hij}{}^k .$$

The expression of $\tilde{\mathcal{R}}_{kji}{}^{h'}$ is not specified as it is not necessary for our purposes.

3.5.2   TWISTED RIEMANNIAN EXTENSIONS DEFINED BY FLAT CONNECTIONS

The case where D is flat is of special interest. We show in this section that both the twisted and modified Riemannian extensions have special significance whenever the base affine connection is flat. We first deal with the twisted Riemannian extension—we refer to [148, 149] for the proofs of the assertions in this section.

Let $g_{D,\phi}$ be a twisted Riemannian extension defined by a flat connection D for arbitrary ϕ. This means that we can choose local coordinates so that

$$g = 2dx_i \circ dx_{i'} + \phi_{ij}(x_1, \ldots, x_n)dx_i \circ dx_j. \tag{3.5}$$

These manifolds have also been called *plane wave* manifolds. First of all, we recall some general facts regarding geodesic completeness. In a coordinate system where the Christoffel symbols vanish, one has

Theorem 3.9. *Let* $\mathcal{M} = (M, g)$ *be as in* Equation (3.5). *Then*

1. *The possibly non-zero components of the curvature tensor and of the covariant derivative* $R^{(\nu)}$ *of the curvature tensor for* $\nu \geq 0$ *are:*

$$R(\partial_{x_i}, \partial_{x_j}, \partial_{x_k}, \partial_{x_\ell}) = \tfrac{1}{2}\{\partial_{x_i}\partial_{x_k}\phi_{j\ell} + \partial_{x_j}\partial_{x_\ell}\phi_{ik} - \partial_{x_i}\partial_{x_\ell}\phi_{jk} - \partial_{x_j}\partial_{x_k}\phi_{i\ell}\},$$
$$R^{(\nu)}(\partial_{x_i}, \partial_{x_j}, \partial_{x_k}, \partial_{x_\ell}; \partial_{x_{j_1}}, \ldots, \partial_{x_{j_\nu}}) = (\partial_{x_{j_1}} \ldots \partial_{x_{j_\nu}})R(\partial_{x_i}, \partial_{x_j}, \partial_{x_k}, \partial_{x_\ell}).$$

2. *All the local Weyl scalar invariants vanish so* $\mathcal{M}$ *is VSI. The Jacobi operator, the skew symmetric curvature operator, and the Szabó operator are all nilpotent operators. These manifolds are Osserman, Ivanov–Petrova, and Szabó. The distribution* $\mathcal{D} := \mathrm{Span}\{\partial_{x_{i'}}\}$ *is a null parallel distribution of maximal dimension giving* $\mathcal{M}$ *the structure of a Walker manifold.*

3. *Assume that* $M = \mathbb{R}^{2n}$. *All geodesics extend for infinite time. The exponential map* $\exp_P$ *from* $T_P M$ *to* M *is a diffeomorphism for any* $P \in M$. *If* $\phi(x) = a_{ij}x_i x_j$ *is a quadratic function, then* $\mathcal{M}$ *is a symmetric space.*

Example 3.10. If $2n \geq 6$, let $\phi := x_1^2 dx_2 \circ dx_2$. Then $\mathcal{J}(\partial_{x_1} \pm \partial_{x_{1'}}) \neq 0$ while $\mathcal{J}(\partial_{x_3} \pm \partial_{x_{3'}}) = 0$. Thus, this manifold is an Osserman symmetric space which is neither timelike Jordan Osserman nor spacelike Jordan Osserman.

Example 3.11. Let $(x, y, z_1, \ldots, z_{n-2}, \tilde{x}, \tilde{y}, \tilde{z}_1, \ldots, \tilde{z}_{n-2})$ be coordinates on $\mathbb{R}^{2n}$. Let $\alpha = \alpha(y)$ be real analytic. Define:

$$\begin{aligned}
\phi_\alpha &:= -2\left\{\alpha(y) + yz_1 + \cdots + y^{n-2}z_{n-2}\right\} dx \otimes dx, \\
g_\alpha &:= \phi_\alpha + 2\{dx \circ d\tilde{x} + dy \circ d\tilde{y} + dz_1 \circ d\tilde{z}_1 + \cdots + dz_{n-2} \circ d\tilde{z}_{n-2}\}.
\end{aligned}$$

Let $\mathcal{M}_\alpha := (\mathbb{R}^{2n}, g_\alpha)$ be the associated plane wave manifold.

Theorem 3.12. *Adopt the notation established above.*

1. *The manifold $\mathcal{M}_{y^2}$ is a symmetric space.*

2. *The manifold $\mathcal{M}_{e^y}$ is a homogeneous space which is curvature modeled on $\mathcal{M}_{y^2}$ and which is not a symmetric space.*

3. *The manifold $\mathcal{M}_{e^y+e^{2y}}$ is:*

    (a) *curvature modeled on the symmetric space $\mathcal{M}_{y^2}$,*

    (b) *$n - 1$ curvature modeled on the homogeneous space $\mathcal{M}_{e^y}$,*

    (c) *$n - 1$ curvature homogeneous but not n curvature homogeneous.*

We can further examine the geometry of these manifolds. Suppose that $\alpha^{(n)} > 0$ and that $\alpha^{(n+1)} > 0$. For $k \geq 1$, define:

$$\sigma_k(\alpha) := \alpha^{(n+k+1)}\{\alpha^{(n)}\}^k\{\alpha^{(n+1)}\}^{-k-1} \ .$$

Theorem 3.13. *Let α_i be real analytic functions with $\alpha_i^{(n)} > 0$ and $\alpha_i^{(n+1)} > 0$. Let $P_i \in \mathbb{R}^{2n}$. The following assertions are equivalent:*

1. *$\sigma_k(\alpha_1)(P_1) = \sigma_k(\alpha_2)(P_2)$ for $k = 1, 2, \ldots.$*

2. *There exists a local isometry $\psi : (\mathbb{R}^{2n}, g_{\alpha_1}, P_1) \to (\mathbb{R}^{2n}, g_{\alpha_2}, P_2)$.*

3. *There exists a global isometry $\psi : (\mathbb{R}^{2n}, g_{\alpha_1}, P_1) \to (\mathbb{R}^{2n}, g_{\alpha_2}, P_2)$.*

This result shows that the numbers $\{\sigma_k(\alpha)(P)\}$ are a complete system of isometry invariants of $(\mathcal{M}_\alpha, P)$. Since α is assumed defined on all of $\mathbb{R}$, Lemma 2.14 shows that $\sigma_1(\alpha)$ is constant if and only if $\alpha^{(n)} = ae^{by}$ for some $a > 0$ and $b > 0$. And furthermore if $\alpha^{(n)} = ae^{by}$, then $\sigma_k(\alpha) = 1$ for all k. Thus, any homogeneous example within this family is isometric to $\mathcal{M}_{e^y}$.

If we set $n = 2$ and omit the $(z_i, \tilde{z}_i)$ variables, we obtain the example

$$g = -2f(y)dx \circ dx + 2\{dx \circ d\tilde{x} + dy \circ d\tilde{y}\} \ .$$

One then has, see [113, 114], that the manifold $\mathcal{M}_{e^y+e^{2y}}$ is a 1 curvature homogeneous complete manifold which is curvature modeled on an irreducible symmetric space and which is not 2 curvature homogeneous.

3.5.3   MODIFIED RIEMANNIAN EXTENSIONS DEFINED BY FLAT CONNECTIONS

Let (M, g, J) be an almost para-Hermitian manifold. Recall that the *para-holomorphic sectional curvature* of a non-degenerate para-holomorphic 2-plane $\pi := \mathrm{Span}\{x, Jx\}$ is given by

$$H(x) = -R(x, Jx, Jx, x)g(x, x)^{-2}.$$

Para-Kaehler manifolds of constant para-holomorphic sectional curvature c are locally symmetric and their curvature tensor [92] is given by $R = \frac{c}{4}\{R_{\mathrm{Id}} + R_J\}$. The importance of modified Riemannian extensions in relation with para-Kaehler geometry is illustrated by the following result [77].

Theorem 3.14. *A para-Kaehler metric of non-zero constant para-holomorphic sectional curvature c is locally isometric to the cotangent bundle of an affine manifold which is equipped with the modified Riemannian extension $g_{D,0,c\,\mathrm{Id,Id}}$ given by Equation (3.3), where D is a flat connection.*

Hence, any para-Kaehler manifold (M, g, J) of non-zero constant para-holomorphic sectional curvature c can be represented as a modified Riemannian extension, in coordinates $(x_1, \ldots, x_n, x_{1'}, \ldots, x_{n'})$, by

$$g = 2dx_i \circ dx_{i'} + c\, x_{i'}x_{j'}dx_i \circ dx_j,$$

where the para-Hermitian structure J is given by $J\partial_{x_i} = \partial_{x_i} - c\, x_{i'}x_{j'}\partial_{x_{j'}}$ and $J\partial_{x_{i'}} = -\partial_{x_{i'}}$.

3.5.4   NILPOTENT WALKER MANIFOLDS

The work of this section is also due to Fiedler [118]. Let $n \geq 3$. We generalize a previous construction given in Section 3.5.2. Take coordinates $(x, u_1, \ldots, u_{n-1}, \tilde{x}, \tilde{u}_1, \ldots, \tilde{u}_{n-1})$ on $\mathbb{R}^{2n}$. We shall suppose that $f = f(u, \tilde{u})$ be a smooth function. We let $\mathcal{M}_f := (\mathbb{R}^{2n}, g_f)$ where

$$g_f := f(u, \tilde{u})dx \circ dx + 2dx \circ d\tilde{x} + 2\sum_{i=1}^{n-1} du_i \circ d\tilde{u}_i.$$

The non-zero Christoffel symbols of the first kind are then given by:

$$g_f(\nabla_{\partial_x}\partial_{u_i}, \partial_x) = g_f(\nabla_{\partial_{u_i}}\partial_x, \partial_x) = -g_f(\nabla_{\partial_x}\partial_x, \partial_{u_i}) = \tfrac{1}{2}\partial_{u_i}f,$$
$$g_f(\nabla_{\partial_x}\partial_{\tilde{u}_i}, \partial_x) = g_f(\nabla_{\partial_{\tilde{u}_i}}\partial_x, \partial_x) = -g_f(\nabla_{\partial_x}\partial_x, \partial_{\tilde{u}_i}) = \tfrac{1}{2}\partial_{\tilde{u}_i}f.$$

Consequently, $\mathcal{D} := \mathrm{Span}\{\partial_{\tilde{x}}, \partial_{\tilde{u}_1}, \ldots, \partial_{\tilde{u}_{n-1}}\}$ is a null parallel distribution and $\mathcal{M}_f$ is a Walker manifold. Note that $\mathcal{M}_f$ is not a Riemannian extension for a general f. However, setting

$$\xi_\ell := -\sum_{i=1}^{\ell}\left\{(u_i + \tilde{u}_i)u_{i+1}^2\right\} - \frac{1}{3}u_1^3, \qquad \eta_\ell := -\sum_{i=1}^{\ell}\left\{2(u_i + \tilde{u}_i)u_{i+1} + u_{i+1}^2\right\} - u_1^2,$$

yields corresponding manifolds $\mathcal{M}_{\xi_\ell}$ and $\mathcal{M}_{\eta_\ell}$ which are twisted Riemannian extensions. More precisely, $\mathcal{M}_{\xi_\ell}$ coincides with the twisted Riemannian extension determined by

$$D_{\partial_x}\partial_x = \frac{1}{2}\sum_{i=1}^{\ell} u_{i+1}^2 \partial_{u_i}, \quad \text{and} \quad \phi = \left(-\sum_{i=1}^{\ell} u_i u_{i+1}^2 - \frac{1}{3}u_1^3\right)dx \otimes dx$$

on $\mathbb{R}^n$ with coordinates $(x, u_1, \ldots, u_{n-1})$. Analogously, $\mathcal{M}_{\eta_\ell}$ corresponds to

$$D_{\partial_x}\partial_x = \sum_{i=1}^{\ell} u_{i+1}\partial_{u_i}, \quad \text{and} \quad \phi = \left(-\sum_{i=1}^{\ell}(2u_i u_{i+1} + u_{i+1}^2) - u_1^2\right)dx \otimes dx$$

on $\mathbb{R}^n$ with coordinates $(x, u_1, \ldots, u_{n-1})$.

Recall that $\mathcal{M}$ is nilpotent Szabó if the associated Szabó operator is nilpotent.

Theorem 3.15.   *Adopt the notation established above.*

1. *The manifold $\mathcal{M}_{\xi_\ell}$ is a Walker manifold of signature (n, n) which is nilpotent Szabó of order $2n - 2$.*

2. *The manifold $\mathcal{M}_{\eta_\ell}$ is a Walker manifold of signature (n, n) which is nilpotent Osserman of order $2n - 2$ and which is nilpotent Ivanov–Petrova of order 3.*

3.5.5   OSSERMAN RIEMANNIAN EXTENSIONS

Recall that $\mathcal{M}$ is *nilpotent Osserman* if the Jacobi operator of $\mathcal{M}$ is nilpotent. We have [129]:

Theorem 3.16.   *Let D be a torsion free connection on M. Let $g_{D,\phi}$ be the twisted Riemannian extension on T^*M given by* Equation (3.4). *The following conditions are equivalent:*

1. *D has nilpotent Jacobi operator.*

2. *$g_{D,\phi}$ is nilpotent Osserman.*

3. *$g_{D,\phi}$ is Osserman.*

Proof. Let $\tilde{X} := \alpha^i \partial_{x_i} + \alpha^{i'}\partial_{x_{i'}}$ be a vector field on T^*M. Then it follows from Lemma 3.8 that the matrix of the Jacobi operator $\tilde{\mathcal{J}}(\tilde{X})$ with respect to the basis $\{\partial_{x_i}, \partial_{x_{i'}}\}$ is of the form

$$\tilde{\mathcal{J}}(\tilde{X}) = \begin{pmatrix} \mathcal{J}(X) & 0 \\ \star & {}^t\mathcal{J}(X) \end{pmatrix}, \tag{3.6}$$

where $\mathcal{J}(X)$ is the matrix of the Jacobi operator corresponding to the vector field $X = \alpha^i \partial_{x_i}$ on M. Now, if (M, D) is assumed to have nilpotent Jacobi operator, then 0 is the only eigenvalue of $\mathcal{J}(X)$. Therefore, it follows from Equation (3.6) that 0 is also the only eigenvalue of $\tilde{\mathcal{J}}(\tilde{X})$ and thus $(T^*M, g_{D,\phi})$ is pseudo-Riemannian Osserman.

Conversely, assume that $(T^*M, g_{D,\phi})$ is pseudo-Riemannian Osserman. If $X := \alpha^i \partial_{x_i}$ is an arbitrary vector field on M, then $\tilde{X} := \alpha^i \partial_{x_i} + \frac{1}{2\alpha^i} \partial_{x_{i'}}$ is a unit vector field at any point of the zero section of T^*M. From Equation (3.6) we see that the characteristic polynomial $p_\lambda(\tilde{\mathcal{J}}(\tilde{X}))$ of $\tilde{\mathcal{J}}(\tilde{X})$ is the square of the characteristic polynomial $p_\lambda(\mathcal{J}(X))$ of $\mathcal{J}(X)$. Since for every unit vector $\tilde{X}$ on T^*M the characteristic polynomial $p_\lambda(\tilde{\mathcal{J}}(\tilde{X}))$ should be the same, we see that for every vector X on M the characteristic polynomial $p_\lambda(\mathcal{J}(X))$ is the same and hence (M, D) has nilpotent Jacobi operator. In particular, all eigenvalues are zero.   □

3.5.6   IVANOV–PETROVA RIEMANNIAN EXTENSIONS

Recall that $\mathcal{M}$ is *nilpotent Ivanov–Petrova* if the associated curvature operator is nilpotent. We have [76]:

Theorem 3.17. *Let D be a torsion free connection on M. Let $g_{D,\phi}$ be the twisted Riemannian extension on T^*M given by* Equation (3.4). *The following conditions are equivalent:*

1. *D has nilpotent curvature operator.*

2. *$g_{D,\phi}$ is nilpotent Ivanov–Petrova.*

3. *$g_{D,\phi}$ is Ivanov–Petrova.*

Proof. Let $\mathcal{R}$ be the curvature operator associated to D and let $\tilde{\mathcal{R}}$ be the corresponding curvature operator associated to the Levi-Civita connection of the metric $g_{D,\phi}$. Let

$$X := \alpha^i \partial_{x_i}, \quad \tilde{X} := \alpha^i \partial_{x_i} + \alpha^{i'} \partial_{x_{i'}},$$
$$Y := \beta^i \partial_{x_i}, \quad \tilde{Y} := \beta^i \partial_{x_i} + \beta^{i'} \partial_{x_{i'}}.$$

We use Lemma 3.8 to express:

$$\tilde{\mathcal{R}}(\tilde{X}, \tilde{Y}) = \begin{pmatrix} \mathcal{R}(X, Y) & 0 \\ \star & -{}^t\mathcal{R}(X, Y) \end{pmatrix}. \tag{3.7}$$

Consequently, we have $\det(\tilde{\mathcal{R}} - \lambda \operatorname{Id}) = \det(\mathcal{R} - \lambda \operatorname{Id}) \cdot \det(-\mathcal{R} - \lambda \operatorname{Id})$ and thus

$$\operatorname{Spec}\{\tilde{\mathcal{R}}(\tilde{X}, \tilde{Y})\} = \operatorname{Spec}\{\mathcal{R}(X, Y)\} \cup \operatorname{Spec}\{-\mathcal{R}(X, Y)\}. \tag{3.8}$$

Suppose Assertion (1) holds so that D has nilpotent curvature operator. It then follows that $\operatorname{Spec}\{\tilde{\mathcal{R}}(\tilde{X}, \tilde{Y})\} = \{0\}$ and hence $g_{D,\phi}$ is nilpotent Ivanov–Petrova. Thus, Assertion (1) implies

Assertion (2); it is immediate that Assertion (2) implies Assertion (3). Suppose finally Assertion (3) holds. Let $\{X, Y\}$ be linearly independent tangent vectors and choose $\{\tilde{X}, \tilde{Y}\}$ an orthonormal basis of a spacelike 2-plane which projects on $\{X, Y\}$. Since $g_{D,\phi}$ is Ivanov–Petrova, we may use Equation (3.8) to conclude that

$$\text{Spec}\{\mathcal{R}(X, Y)\} \cup \text{Spec}\{-\mathcal{R}(X, Y)\}$$

is independent of the choice of X and Y. Let α be a non-zero real number. Since

$$\begin{aligned}
&\text{Spec}\{\mathcal{R}(\alpha X, Y)\} \cup \text{Spec}\{-\mathcal{R}(\alpha X, Y)\} \\
=~&\alpha\{\text{Spec}\{\mathcal{R}(X, Y)\} \cup \text{Spec}\{-\mathcal{R}(X, Y)\}\}\,,
\end{aligned}$$

we conclude $\text{Spec}\{\mathcal{R}(X, Y)\} = \{0\}$ and hence D has nilpotent curvature operator. $\square$

CHAPTER 4

Three-Dimensional Lorentzian Walker Manifolds

4.1   INTRODUCTION

In this Chapter, we discuss the geometry of 3 dimensional Walker manifolds by showing both some of their specific features as well as their influence on the study of 3 dimensional Lorentzian geometry. From the point of view of Walker geometry, 3 dimensional Walker manifolds are the first non-trivial case for consideration. Moreover, the fact that all geometric data is encoded in a single function makes this geometry more tractable than the corresponding higher dimensional cases.

We begin in Section 4.2 with some historical remarks. Then in Section 4.3, we introduce 3 dimensional Walker geometry. We give a canonical form, locally, for the metric in Section 4.3.1 and analyze the possible Jordan normal forms for the Ricci operator in Section 4.3.2. The Christoffel symbols and the curvature tensor as well as some of the curvature operators are discussed in Section 4.3.3. A classification of locally symmetric 3 dimensional Walker manifolds is given in Section 4.3.4. Finally, we study some generalizations of Einstein manifolds in Section 4.3.5 and several properties of the curvature operators in Section 4.3.6.

In Section 4.4, we discuss a number of curvature conditions and treat foliated Walker manifolds and contact Walker manifolds. In Section 4.5, we investigate strict Walker geometry in dimension 3. In Section 4.6, we return to the general setting and analyze 3 dimensional homogeneous and symmetric Lorentzian manifolds. Finally, in Section 4.7, we discuss curvature homogeneous Lorentzian manifolds; Walker geometry once again plays a central role.

4.2   HISTORY

Three dimensional geometry plays a central role in the investigation of many problems in Riemannian and Lorentzian geometry. The fact that the Ricci operator completely determines the curvature tensor is crucial to these investigations.

Einstein manifolds of dimension 3 are of constant sectional curvature and 3 dimensional manifolds with parallel Ricci operator are locally symmetric. Gray [162] considered the space of covariant derivatives of Ricci tensors to introduce some generalizations of the Einstein condition. The three primitive classes of manifolds, consisting on those with cyclic parallel Ricci tensor, Codazzi Ricci tensor and $\mathcal{C}^{\perp}$ have been extensively studied in the Riemannian case. Riemannian manifolds of dimension 3 with cyclic parallel Ricci tensor are locally homogeneous naturally reductive [222]. $\mathcal{C}^{\perp}$ Riemannian 3 dimensional manifolds (see Section 4.3.5) are classified by Berndt [25] while the complete classification of 3 dimensional manifolds with Codazzi Ricci tensor is still open.

In showing that any Riemannian manifold of dimension 3 with cyclic parallel Ricci operator is locally homogeneous one makes use of the theory of curvature homogeneous Riemannian manifolds which was introduced in Section 2.5.4. Any locally homogeneous pseudo-Riemannian manifold is clearly curvature homogeneous of any order. It is well-known that 0 curvature homogeneity does not suffice to ensure local homogeneity in dimension greater than 2 as pointed out by Takagi [249], but 3 dimensional Riemannian manifolds which are 1 curvature homogeneous are locally homogeneous [232]. The situation is more complicated in the Lorentzian case (which is partially due to the different Jordan normal forms of the Ricci operator in signature $(1, 2)$) and 1 curvature homogeneity does not guarantee local homogeneity [65]. A complete description of 1 curvature homogeneous Lorentzian 3 dimensional manifolds is given in [64] showing that their Ricci operators must be diagonalizable or otherwise there is a double root of their minimal polynomial. The class of 0 curvature homogeneous 3 dimensional manifolds is much larger and all possible Jordan normal forms are available for the Ricci operator as shown in [72].

Curvature homogeneity (and thus local homogeneity) can be characterized in the Riemannian setting by using scalar curvature invariants [226]. In fact, locally homogeneous Riemannian manifolds can be detected by the constancy of their scalar curvature invariants [226]. This fact (which relies on the compactness of the structure group) does not extend to the pseudo-Riemannian case, where manifolds exist all whose scalar curvature invariants vanish (VSI) but they are not locally homogeneous as shown in [65, 80, 183, 225]. Walker metrics play a distinguished role in understanding many of these problems. Other related references are [19, 63, 66, 68, 69, 70, 71, 170].

4.3   THREE DIMENSIONAL WALKER GEOMETRY

Let $\mathfrak{M} := (V, \langle \cdot, \cdot \rangle, A)$ be a curvature model on a 3 dimensional vector space as discussed in Section 1.3.3 and let ρ be the associated Ricci tensor given in Equation (1.6). Since the Weyl curvature tensor vanishes identically if $m = 3$, the curvature tensor of a 3 dimensional pseudo-Riemannian manifold is completely determined by the Ricci tensor. This fact is essential in the development of the present Chapter.

4.3.1   ADAPTED COORDINATES

The Walker coordinates introduced in Section 3.3 simplify when the null r-plane $\mathcal{D}$ has maximum dimensionality. Since dim $\mathcal{D} = r \leq \frac{1}{2}m$ two cases occur depending on whether m is odd or even. Since our objective in this Chapter is the analysis of 3 dimensional Walker manifolds, we begin with the following general result [255]:

Theorem 4.1.   *A canonical form for a $2r + 1$ dimensional pseudo-Riemannian manifold $\mathcal{M}$ admitting a parallel field of null r dimensional planes $\mathcal{D}$ is given by the metric tensor in matrix form as*

$$(g_{ij}) = \begin{pmatrix} 0 & 0 & \mathrm{Id}_r \\ 0 & \varepsilon & 0 \\ \mathrm{Id}_r & 0 & B \end{pmatrix}$$

where Id_r *is the* $r \times r$ *identity matrix,* B *is a symmetric* $r \times r$ *matrix whose entries are functions of the coordinates* $(x_1, \ldots, x_{2r+1})$ *and* $\varepsilon = \pm 1$.

Theorem 3.2 implies that the null plane is strictly parallel if and only if the entries of B in Theorem 4.1 are independent of the coordinates $(x_1, \ldots, x_r)$ [255].

We clear the previous notation. Throughout Chapter 4 we shall consider the manifolds $\mathcal{M}_f$ which are defined as follows. Let $f = f(x_1, x_2, x_3)$ be a smooth function which is defined on an open subset $\mathcal{O}$ of $\mathbb{R}^3$. Let $\varepsilon = \pm 1$. We define $\mathcal{M}_f := (\mathcal{O}, g_f)$ where

$$g_f := 2dx_1 \circ dx_3 + \varepsilon dx_2 \circ dx_2 + f(x_1, x_2, x_3)dx_3 \circ dx_3 . \tag{4.1}$$

The manifold $\mathcal{M}_f$ has signature $(1, 2)$ if $\varepsilon = +1$; $\mathcal{M}_f$ has signature $(2, 1)$ if $\varepsilon = -1$. Let $\mathcal{M}$ be a 3 dimensional Walker manifold. We use Theorem 4.1 to see $\mathcal{M}$ is locally isometric to $\mathcal{M}_f$ for some suitably chosen $\{f, \mathcal{O}, \varepsilon\}$. We note that the Walker coordinate system is not unique and thus f is not determined uniquely; in certain cases further normalizations of f are possible – this will play a role in our subsequent discussions. In the remainder of Section 4.3, we shall summarize basic results concerning the manifolds $\mathcal{M}_f$ and refer to [80] and the references therein for details.

4.3.2 THE JORDAN NORMAL FORM OF THE RICCI OPERATOR

In the Riemannian setting, there always exists an orthonormal basis diagonalizing the Ricci operator ρ. However, in the Lorentzian case, the Ricci operator need not be diagonalizable even though it is self-adjoint. Let the *Ricci curvatures* be the (possibly complex) eigenvalues of the Ricci operator ρ. There are four different cases which occur according to the behavior of the Jordan normal form of the Ricci operator:

(Ia). The Ricci operator is diagonalizable, i.e.,

$$\rho = \begin{pmatrix} \alpha & 0 & 0 \\ 0 & \beta & 0 \\ 0 & 0 & \gamma \end{pmatrix}.$$

(Ib). The Ricci operator has complex eigenvalues, i.e.,

$$\rho = \begin{pmatrix} \alpha & -\beta & 0 \\ \beta & \alpha & 0 \\ 0 & 0 & \gamma \end{pmatrix} \quad \text{where} \quad \beta \neq 0.$$

(II). There is a 2×2 Jordan block and hence a double root of the minimal polynomial of ρ, i.e.,

$$\rho = \begin{pmatrix} \alpha & 0 & 0 \\ 0 & \beta & 0 \\ 0 & 1 & \beta \end{pmatrix}.$$

(III). There is a 3×3 Jordan block and hence a triple root of the minimal polynomial of ρ, i.e.,

$$\rho = \begin{pmatrix} \alpha & 0 & 0 \\ 1 & \alpha & 0 \\ 0 & 1 & \alpha \end{pmatrix}.$$

This categorization plays a central role in what follows.

4.3.3   CHRISTOFFEL SYMBOLS, CURVATURE, AND THE RICCI TENSOR

We begin our investigation by studying the Levi-Civita connection of $\mathcal{M}_f$.

Lemma 4.2.   *Let* $\mathcal{M}_f$ *be as defined in* Equation (4.1).

1. *The non-zero components of the Christoffel symbols are:*

$$\nabla_{\partial_{x_1}}\partial_{x_3} = \tfrac{1}{2}f_1\partial_{x_1}, \quad \nabla_{\partial_{x_2}}\partial_{x_3} = \tfrac{1}{2}f_2\partial_{x_1}, \quad \nabla_{\partial_{x_3}}\partial_{x_3} = \tfrac{1}{2}(ff_1 + f_3)\partial_{x_1} - \tfrac{\varepsilon}{2}f_2\partial_{x_2} - \tfrac{1}{2}f_1\partial_{x_3}\,.$$

2. *The distribution* $\mathcal{D} := \mathrm{Span}\{\partial_{x_1}\}$ *is a null parallel distribution. The distribution* $\mathcal{D}$ *admits a parallel spanning vector field if and only if* $f(x_1, x_2, x_3) = a(x_2, x_3) + x_1 b(x_3)$.

Proof. The calculation of the Christoffel symbols is immediate. It now follows that $\mathcal{D}$ is a null parallel distribution. Suppose that $s = e^\phi \partial_{x_1}$ is a parallel vector field. Then ϕ must satisfy the equations $\partial_{x_3}\phi = -\tfrac{1}{2}f_1$, $\partial_{x_1}\phi = 0$, $\partial_{x_2}\phi = 0$. It now follows that $f_{11} = f_{12} = 0$.   □

Remark 4.3.   If $\mathcal{D}$ admits a parallel spanning vector field, we can renormalize the Walker coordinates so that $f = a(x_2, x_3)$ is independent of the parameter x_1 and so $\mathcal{D} = \mathrm{Span}\{\partial_{x_1}\}$. A more detailed treatment of this specific case is carried out in Section 4.5.

We use Lemma 4.2 to establish:

Lemma 4.4.

1. *The non-zero components of the curvature tensor of* $\mathcal{M}_f$ *are*

$$\mathcal{R}(\partial_{x_1}, \partial_{x_3})\partial_{x_1} = \tfrac{1}{2}f_{11}\partial_{x_1}, \quad \mathcal{R}(\partial_{x_1}, \partial_{x_3})\partial_{x_3} = \tfrac{1}{2}ff_{11}\partial_{x_1} - \tfrac{\varepsilon}{2}f_{12}\partial_{x_2} - \tfrac{1}{2}f_{11}\partial_{x_3},$$
$$\mathcal{R}(\partial_{x_1}, \partial_{x_3})\partial_{x_2} = \tfrac{1}{2}f_{12}\partial_{x_1}, \quad \mathcal{R}(\partial_{x_2}, \partial_{x_3})\partial_{x_1} = \tfrac{1}{2}f_{12}\partial_{x_1},$$
$$\mathcal{R}(\partial_{x_2}, \partial_{x_3})\partial_{x_2} = \tfrac{1}{2}f_{22}\partial_{x_1}, \quad \mathcal{R}(\partial_{x_2}, \partial_{x_3})\partial_{x_3} = \tfrac{1}{2}ff_{12}\partial_{x_1} - \tfrac{\varepsilon}{2}f_{22}\partial_{x_2} - \tfrac{1}{2}f_{12}\partial_{x_3}.$$

2. *The Ricci tensor of* $\mathcal{M}_f$ *is* $f_{11}\,dx_1 \circ dx_3 + f_{12}\,dx_2 \circ dx_3 + \tfrac{1}{2}(ff_{11} - \varepsilon f_{22})\,dx_3 \circ dx_3\,.$

3. *The Ricci operator ρ of $\mathcal{M}_f$ is*

$$\rho = \begin{pmatrix} \frac{1}{2}f_{11} & \frac{1}{2}f_{12} & -\frac{\varepsilon}{2}f_{22} \\ 0 & 0 & \frac{\varepsilon}{2}f_{12} \\ 0 & 0 & \frac{1}{2}f_{11} \end{pmatrix}.$$

4. *The Ricci operator of $\mathcal{M}_f$ has eigenvalues: $\lambda_1 = 0, \lambda_2 = \lambda_3 = \frac{1}{2}f_{11}; \tau = f_{11}$.*

5. *$\mathcal{M}_f$ is Einstein if and only if $\mathcal{M}_f$ is flat.*

6. *$\mathcal{M}_f$ has constant Ricci curvatures if and only if $f = \kappa x_1^2 + x_1 P(x_2, x_3) + \xi(x_2, x_3)$.*

7. *If $\mathcal{M}_f$ has constant Ricci curvatures $\{0, \kappa, \kappa\}, \kappa \neq 0$, then ρ is diagonalizable if and only if $f(x_1, x_2, x_3) = \kappa x_1^2 + x_1(P(x_3) + x_2 Q(x_3)) + x_2(\frac{1}{4\kappa}x_2 Q(x_3)^2 + T(x_3)) + \xi(x_3)$.*

Remark 4.5. $\mathcal{M}_f$ has a 2-step nilpotent Ricci operator if and only if $f_{11} = f_{12} = 0$. Moreover, in such a case $\mathcal{D}$ admits a parallel spanning vector field and the metric can be renormalized to be strict (see Theorem 3.2).

4.3.4   LOCALLY SYMMETRIC WALKER MANIFOLDS

We devote this section to the analysis of local symmetry in 3 dimensional Walker manifolds. We have the following observation that follows from the results of Section 4.3.3:

Theorem 4.6. *$\mathcal{M}_f$ is locally symmetric if and only if f has one of the following forms:*

1. $f(x_1, x_2, x_3) = x_1^2 \kappa + x_1(x_2 P(x_3) + Q(x_3)) + \frac{x_2^2}{4\kappa} P(x_3)^2 + \frac{x_2}{\kappa}(P'(x_3)$
   $+ \frac{1}{2} P(x_3) Q(x_3)) + \xi(x_3)$ *for any real constant $\kappa \neq 0$.*

2. $f(x_1, x_2, x_3) = x_1 Q(x_3) + x_2^2 \alpha(x_3) + x_2 \beta(x_3) + \xi(x_3)$ *where $\alpha' + Q\alpha = 0$.*

Remark 4.7. Let $\mathcal{M}_f$ be as in Theorem 4.6 (1). Then $\tau = 2\kappa$ and

$$\rho = \begin{pmatrix} \kappa & \frac{1}{2}P(x_3) & -\frac{\varepsilon}{4\kappa}P(x_3)^2 \\ 0 & 0 & \frac{\varepsilon}{2}P(x_3) \\ 0 & 0 & \kappa \end{pmatrix}.$$

This is always diagonalizable, so $\mathcal{M}_f$ is locally a product of a Lorentzian surface of constant scalar curvature κ (defined by the coordinates (x_1, x_3)) and an interval.

Remark 4.8.   Let $\mathcal{M}_f$ be as in Theorem 4.6 (2). Then $\tau = 0$ and

$$\rho = \begin{pmatrix} 0 & 0 & -\varepsilon\alpha(x_3) \\ 0 & 0 & 0 \\ 0 & 0 & 0 \end{pmatrix}.$$

This is diagonalizable if and only if $\alpha = 0$, i.e., the metric is flat.

4.3.5   EINSTEIN-LIKE MANIFOLDS

Einstein metrics as well as constant scalar curvature metrics are important classes of pseudo-Riemannian manifolds. Every Einstein metric has parallel Ricci tensor, however, not every manifold with constant scalar curvature has parallel Ricci tensor. Thus, one can say that manifolds with parallel Ricci tensor lie between both classes above.

A. Gray [162] generalized the parallel Ricci condition in two different ways. A pseudo-Riemannian manifold $\mathcal{M}$ is said to have *cyclic parallel Ricci tensor* if

$$(\nabla_X \rho)(X, X) = 0$$

for any vector field X or, equivalently, if for all X, Y, and Z one has the identity:

$$(\nabla_X \rho)(Y, Z) + (\nabla_Y \rho)(Z, X) + (\nabla_Z \rho)(X, Y) = 0.$$

Similarly, the Ricci tensor of $\mathcal{M}$ is said to be *Codazzi* if for all X, Y, and Z one has

$$(\nabla_X \rho)(Y, Z) = (\nabla_Y \rho)(X, Z).$$

Either of these two conditions implies that $\mathcal{M}$ has constant scalar curvature.

Translating the symmetries of the covariant derivative of the Ricci tensor to an algebraic context, Gray considered the space $\mathfrak{S}$ of 3 tensors with the symmetries of $(\nabla.\rho)(\cdot, \cdot)$. Suppose given a curvature model $\mathfrak{M} = (V, \langle \cdot, \cdot \rangle, A)$. Let $\{e_1, \ldots, e_m\}$ be a basis for V. Let $\varepsilon_{ij} := \langle e_i, e_j \rangle$. Define

$$\mathfrak{S} := \{\sigma \in \otimes^3 V^* : \sigma(x, y, z) = \sigma(x, z, y) \text{ and } \varepsilon^{ij}\sigma(x, e_i, e_j)$$
$$= 2\varepsilon^{ij}\sigma(e_i, e_j, x) \,\forall x, y, z \in V\}.$$

Give $\mathfrak{S}$ the inner product $\langle \sigma_1, \sigma_2 \rangle = \varepsilon^{i_1 j_1}\varepsilon^{i_2 j_2}\varepsilon^{i_3 j_3}\sigma_1(e_{i_1}, e_{i_2}, e_{i_3})\sigma_2(e_{j_1}, e_{j_2}, e_{j_3})$. We introduce the following subspace which, in the geometric context, would correspond to manifolds of constant scalar curvature:

$$\mathcal{C} := \{\sigma \in \mathfrak{S} : \varepsilon^{ij}\sigma(x, e_i, e_j) = 0\}.$$

We also define the following two subspaces of $\mathcal{C}$:

$$\mathcal{A} := \{\sigma \in \mathfrak{S} : \sigma(x, y, z) + \sigma(y, z, x) + \sigma(z, x, y) = 0 \,\forall x, y, z \in V\}, \quad \text{and}$$
$$\mathcal{B} := \{\sigma \in \mathfrak{S} : \sigma(x, y, z) = \sigma(y, x, z) \,\forall x, y, z \in V\}.$$

Note that the covariant derivative of the Ricci tensor of manifolds with cyclic parallel Ricci tensors at any point lies in $\mathcal{A}$; whereas the covariant derivative of the Ricci tensor of manifolds with Codazzi Ricci tensor at any point lies in $\mathcal{B}$. One has the following orthogonal direct sum decomposition [162]:

$$\mathfrak{S} = \mathcal{A} \oplus \mathcal{B} \oplus \mathcal{C}^{\perp} .$$

The two conditions given above, namely cyclic parallel and Codazzi Ricci tensors, arise in a natural way from the decomposition of the space of 3 tensors verifying the symmetries of $(\nabla. \rho)(\cdot, \cdot)$, and decompose orthogonally the space $\mathcal{C}$. Manifolds corresponding to models in $\mathcal{C}^{\perp}$ are characterized by the following identity which must be satisfied for all vector fields X, Y, and Z:

$$(\nabla_X \rho)(Y, Z) = \tfrac{1}{(m+2)(m-1)} \left\{ m X(\tau) \langle Y, Z \rangle + \tfrac{m-2}{2} (Y(\tau) \langle X, Z \rangle + Z(\tau) \langle X, Y \rangle) \right\} . \qquad (4.2)$$

Clearly, any manifold with parallel Ricci tensor has cyclic parallel Ricci tensor and Codazzi Ricci tensor, and satisfies the Equation (4.2), but the converse is not true in general. However, any manifold satisfying any two of these three conditions necessarily has parallel Ricci tensor.

Theorem 4.9.

1. *The Ricci tensor of $\mathcal{M}_f$ is cyclic parallel if and only if it is parallel.*

2. *$\mathcal{M}_f$ is a $\mathcal{C}^{\perp}$ manifold if and only if the Ricci tensor of $\mathcal{M}_f$ is parallel.*

Recall from Section 1.3.6 that the Schouten tensor C is given by:

$$C = \frac{1}{m-2} \left(\rho - \frac{\tau}{2(m-1)} g \right) .$$

A 3 dimensional manifold is locally conformally flat if and only if the Schouten tensor C is Codazzi, i.e.,

$$c(X, Y, Z) := (\nabla_X C)(Y, Z) - (\nabla_Y C)(X, Z) = 0 .$$

Lemma 4.10.   *The non-zero components of the Codazzi equations for the Schouten tensor of the manifold $\mathcal{M}_f$ are:*

$$c(\partial_{x_2}, \partial_{x_1}, \partial_{x_2}) = -c(\partial_{x_1}, \partial_{x_2}, \partial_{x_2}) = \tfrac{\varepsilon}{2} f_{111},$$

$$c(\partial_{x_3}, \partial_{x_1}, \partial_{x_2}) = -c(\partial_{x_1}, \partial_{x_2}, \partial_{x_3}) = c(\partial_{x_2}, \partial_{x_1}, \partial_{x_3}) = -c(\partial_{x_1}, \partial_{x_3}, \partial_{x_2}) = -\tfrac{1}{2} f_{112},$$

$$c(\partial_{x_3}, \partial_{x_1}, \partial_{x_3}) = -c(\partial_{x_1}, \partial_{x_3}, \partial_{x_3}) = \tfrac{\varepsilon}{2} f_{122},$$

$$c(\partial_{x_3}, \partial_{x_2}, \partial_{x_2}) = -c(\partial_{x_2}, \partial_{x_3}, \partial_{x_2}) = -\tfrac{1}{2}(f_{122} + \varepsilon f_{113}),$$

$$c(\partial_{x_3}, \partial_{x_2}, \partial_{x_3}) = -c(\partial_{x_2}, \partial_{x_3}, \partial_{x_3}) = \tfrac{1}{4}(2\varepsilon f_{222} + f_1 f_{12} + 2f_{123} - f_2 f_{11}).$$

Note that the Ricci tensor of any locally conformally flat manifold with constant scalar curvature is Codazzi. By Lemma 4.10, $\mathcal{M}_f$ is locally conformally flat if and only if

$$f(x_1, x_2, x_3) = x_1^2 \kappa + x_1(x_2 P(x_3) + Q(x_3)) + \xi(x_2, x_3) \quad \text{where}$$
$$x_2 P^2 + PQ + 2P' - 2\kappa \xi_2 + 2\varepsilon \xi_{222} = 0.$$

Theorem 4.11. *The following assertions are equivalent:*

1. *The Ricci tensor of $\mathcal{M}_f$ is Codazzi.*

2. *$\mathcal{M}_f$ is locally conformally flat.*

3. *We have $f(x_1, x_2, x_3) = x_1^2 \kappa + x_1(x_2 P(x_3) + Q(x_3)) + \xi(x_2, x_3)$ where*

$$2\xi_{22}(x_2, x_3) - 2\varepsilon \kappa \xi(x_2, x_3) = \gamma(x_3) - \tfrac{x_2^2}{2}\varepsilon P(x_3)^2 - \varepsilon x_2(P(x_3)Q(x_3) + 2P'(x_3)).$$

Remark 4.12. It follows that if $\mathcal{M}_f$ is locally conformally flat and has vanishing scalar curvature (i.e., $\kappa = 0$), then $f = f(x_1, x_2, x_3)$ satisfies

$$\begin{aligned}
f &= x_1(x_2 P(x_3) + Q(x_3)) - \varepsilon \tfrac{1}{48} x_2^4 P(x_3)^2 - \varepsilon \tfrac{1}{12} x_2^3 (P(x_3)Q(x_3) + 2P'(x_3)) \\
&+ \tfrac{1}{4} x_2^2 \gamma(x_3) + x_2 \eta(x_3) + \delta(x_3).
\end{aligned}$$

4.3.6   THE SPECTRAL GEOMETRY OF THE CURVATURE TENSOR

The 3 dimensional case is an exceptional case in the study of Ivanov–Petrova manifolds, where the problem is completely solved at the algebraic level in the Riemannian setting but not at the differentiable level [172, 207], where the problem becomes more complicated. Since in dimension 3 the Ricci tensor determines the curvature tensor, the investigation of Ivanov–Petrova manifolds is translated into the problem of classifying metrics with prescribed Ricci tensor. A complete solution to the 3 dimensional Ivanov–Petrova algebraic problem in Lorentzian signature is given in [127] as follows.

Theorem 4.13. *Let $\mathfrak{M} := (V, \langle \cdot, \cdot \rangle, A)$ be a 3 dimensional Lorentzian curvature model. $\mathfrak{M}$ is Ivanov–Petrova if and only if one of the following holds:*

1. *$\mathfrak{M}$ has constant sectional curvature.*

2. The Ricci operator ρ is diagonalizable with eigenvalues $\{0, 0, \kappa\}$, $\kappa \neq 0$.

3. The Ricci operator ρ is a 2-step nilpotent operator.

Note that case (3) does not occur in the Riemannian setting [171, 172]. The Ricci operator of any 3 dimensional Walker manifold does not permit the eigenvalue structure given by (2) in Theorem 4.13. Then, as a consequence of Remark 4.5, one has the following geometric consequence:

Theorem 4.14.   *The following assertions are equivalent conditions:*

1. $\mathcal{M}_f$ is Ivanov–Petrova.

2. $f(x_1, x_2, x_3) = a(x_2, x_3) + x_1 b(x_3)$.

3. The distribution $\mathcal{D}$ admits a parallel spanning vector field.

Remark 4.15.   By Remark 4.3, if condition (3) holds in Theorem 4.14, then $\mathcal{M}_f$ is a strict Walker manifold.

4.3.7   CURVATURE COMMUTATIVITY PROPERTIES

We recall the definitions of Section 1.4.5. Commutativity properties of curvature operators have been investigated in the 3 dimensional setting in [125], where the following algebraic descriptions are obtained.

Theorem 4.16.   *Let $\mathfrak{M} := (V, \langle \cdot, \cdot \rangle, A)$ be a 3 dimensional Lorentzian curvature model.*

1. $\mathfrak{M}$ is curvature–Ricci commuting if and only if one of the following holds:

   (a) $\mathfrak{M}$ is of constant sectional curvature.

   (b) The Ricci operator ρ is diagonalizable with eigenvalues $\{0, \kappa, \kappa\}$, $\kappa \neq 0$.

   (c) The Ricci operator ρ is a 2-step nilpotent operator.

2. $\mathfrak{M}$ is curvature–curvature commuting if and only if one of the following holds:

   (a) A is identically zero.

   (b) The Ricci operator ρ is diagonalizable with eigenvalues $\{0, \kappa, \kappa\}$, $\kappa \neq 0$.

   (c) The Ricci operator ρ is a 2-step nilpotent operator.

These results in the algebraic setting then give rise to the following geometric consequence:

Theorem 4.17.   *The following conditions are equivalent conditions:*

1. $\mathcal{M}_f$ is curvature–Ricci commuting.

2. $\mathcal{M}_f$ is curvature–curvature commuting.

3. One of the following holds:

   (a) The Ricci operator is diagonalizable, i.e., $f_{12}^2 - f_{11}f_{22} = 0$.

   (b) The Ricci operator satisfies $\rho^2 = 0$, i.e., $f(x_1, x_2, x_3) = a(x_2, x_3) + x_1 b(x_3)$ or, equivalently, the distribution $\mathcal{D}$ admits a parallel spanning vector field.

4.4   LOCAL GEOMETRY OF WALKER MANIFOLDS WITH $\tau \neq 0$

In this section, we report work of [126]. Recall from Lemma 4.4 that the Ricci operator of a 3 dimensional Walker manifold has eigenvalues $\lambda_1 = 0$, $\lambda_2 = \lambda_3 = \frac{1}{2}\tau$. In this section, we assume $\tau \neq 0$ at each point of $\mathcal{M}$. The eigenvalue $\lambda_1 = 0$ is a distinguished eigenvalue. We denote the associated eigenvector by

$$N = -\frac{f_{12}}{\tau}\partial_{x_1} + \partial_{x_2}.$$

Assume $\varepsilon = 1$ so N is spacelike. Let $\mathcal{N}$ be the associated flow. Let ω be the associated 1 form:

$$\omega = dx_2 - \frac{f_{12}}{\tau}dx_3 \ .$$

We then have that

$$\omega \wedge d\omega = \left(\frac{f_{12}}{\tau}\right)_1 dx_1 \wedge dx_2 \wedge dx_3.$$

This distinguishes two natural subclasses:

1. If $(\frac{f_{12}}{\tau})_1 = 0$, then the Ricci operator defines a foliation. For instance, this is the case whenever the scalar curvature is a non-zero constant.

2. If $(\frac{f_{12}}{\tau})_1 \neq 0$ at each point, then the Ricci operator defines a contact structure.

   We study these two classes of 3 dimensional Walker manifolds separately.

4.4.1   FOLIATED WALKER MANIFOLDS

Suppose that the scalar curvature τ of $\mathcal{M}_f$ never vanishes. Decompose $TM = \mathcal{N} \oplus \mathcal{N}^\perp$ where $\mathcal{N}$ is the 1 dimensional distribution given by the eigenspace associated to the distinguished eigenvalue of the Ricci operator and $\mathcal{N}^{\perp}$ is the orthogonal complement. The *second fundamental form L of $\mathcal{N}$* and $L^{\perp}$ of $\mathcal{N}^{\perp}$ are given [228], respectively, by:

$$\langle L_N N, X \rangle = \langle \nabla_N N, X \rangle, \qquad\qquad \langle L_N X, N \rangle = -\langle L_N N, X \rangle,$$

$$L_X = 0, \qquad\qquad\qquad\qquad \langle L_N N, N \rangle = \langle L_N X, Y \rangle = 0,$$

$$\langle L_X^{\perp} Y, N \rangle = \tfrac{1}{2} \langle \nabla_X Y + \nabla_Y X, N \rangle, \quad \langle L_X^{\perp} N, Y \rangle = -\langle L_X^{\perp} Y, N \rangle,$$

$$L_N^{\perp} = 0, \qquad\qquad\qquad\qquad \langle L_X^{\perp} Y, Z \rangle = \langle L_X^{\perp} N, N \rangle = 0.$$

The distribution $\mathcal{N}^{\perp}$ has signature $(1, 1)$. We take the orthonormal frame :

$$Z_+ = \tfrac{1}{2} \left(1 - \left(\tfrac{f_{12}}{\tau} \right)^2 - f \right) \partial_{x_1} + \tfrac{f_{12}}{\tau} \partial_{x_2} + \partial_{x_3},$$

$$Z_- = -\tfrac{1}{2} \left(1 + \left(\tfrac{f_{12}}{\tau} \right)^2 + f \right) \partial_{x_1} + \tfrac{f_{12}}{\tau} \partial_{x_2} + \partial_{x_3}.$$

Recall that a plane field is said to be *totally geodesic* if its second fundamental form vanishes and *minimal* if its corresponding second fundamental form is trace free. The integrability of the normal plane field can be now stated as follows:

Theorem 4.18. *Suppose τ is nowhere vanishing. The following assertions are equivalent:*

1. *The plane field $\mathcal{N}^{\perp}$ is integrable.*

2. *N is a divergence free locally defined vector field on $\mathcal{M}_f$.*

3. *$\mathcal{N}^{\perp}$ is a minimal plane field on $\mathcal{M}_f$ (i.e., $\mathrm{Tr}\, L^{\perp} = 0$).*

Remark 4.19. It follows from the previous theorem that the shape operator $S(N)$ of the distribution $\mathcal{N}^{\perp}$, given by $\langle S(N)X, Y \rangle = \langle L_X^{\perp} N, Y \rangle$, has eigenvalues $\alpha_1 = 0$ and $\alpha_2 = -(\tfrac{f_{12}}{\tau})_1$. Hence, $S(N)$ is nilpotent whenever $\mathcal{N}^{\perp}$ is integrable.

Now, if $\mathcal{N}^{\perp}$ is integrable, then a straightforward calculation using the Gauss equation shows that the sectional curvature $K^{\mathcal{N}^{\perp}}$ of the (Lorentzian) leafs of $\mathcal{N}^{\perp}$ satisfies

$$K^{\mathcal{N}^{\perp}} = -R^{\mathcal{N}^{\perp}}(Z_+, Z_-, Z_-, Z_+)$$

$$= -R(Z_+, Z_-, Z_-, Z_+)$$

$$\quad - \langle S(N)Z_+, S(N)Z_+ \rangle \langle S(N)Z_-, Z_- \rangle + \langle S(N)Z_+, S(N)Z_- \rangle^2$$

$$= \tfrac{1}{2} \tau.$$

Thus, if τ never vanishes and if $(\tfrac{f_{12}}{\tau})_1 = 0$, then $\mathcal{M}_f$ is foliated by Lorentzian surfaces whose Gaussian curvature equals one half of the scalar curvature of the Walker manifold.

A foliation $\mathcal{F}$ is said to be *bundle-like* if it is locally defined by a Riemannian submersion or, equivalently, if

$$(\mathcal{L}_{\mathcal{F}}g)(\mathcal{F}^{\perp}, \mathcal{F}^{\perp}) = 0$$

where $\mathcal{L}$ denotes the Lie derivative. Further note that if $\mathcal{F}$ is bundle-like, then the orthogonal plane field $\mathcal{F}^{\perp}$ is totally geodesic (although not necessarily integrable).

Theorem 4.20.   *Let τ never vanish.*

1. *The following assertions are equivalent:*

   (a) $\mathcal{N}$ *defines a bundle-like foliation of $\mathcal{M}_f$.*

   (b) $\mathcal{N}^{\perp}$ *is a totally geodesic distribution.*

   (c) $\frac{1}{2}f_2 - \frac{1}{2}f_1\frac{f_{12}}{\tau} - (\frac{f_{12}}{\tau})_3 - \frac{f_{12}}{\tau}(\frac{f_{12}}{\tau})_2 = 0.$

2. *Assume in addition that the plane field $\mathcal{N}^{\perp}$ is integrable. The following assertions are equivalent:*

   (a) $\mathcal{N}^{\perp}$ *defines a bundle-like foliation.*

   (b) N *is a geodesic vector field.*

   (c) $(\frac{f_{12}}{\tau})_1 = (\frac{f_{12}}{\tau})_2 = 0.$

Remark 4.21.   Observe that although N is not a geodesic vector field in general i.e., we may have that $\nabla_N N \neq 0$), N is isotropically geodesic since $|\nabla_N N|^2 = 0$.

A special case where Theorem 4.18 applies is that of Walker manifolds of constant scalar curvature (since $\tau = f_{11}$). Hence, as a consequence of Theorem 4.18 and Remark 4.19, *any 3 dimensional Walker manifold of constant non-zero scalar curvature is foliated by minimal surfaces of constant scalar curvature.*

Theorem 4.22.   *The Ricci foliation $\mathcal{N}^{\perp}$ of $\mathcal{M}_f$ is totally geodesic if and only if*

$$f(x_1, x_2, x_3) = \frac{\kappa}{2}x_1^2 + x_1 a(x_2, x_3) + \frac{1}{2\kappa}a(x_2, x_3)^2 + \frac{2}{\kappa}a_3(x_2, x_3) + \frac{1}{\kappa^2}a_2(x_2, x_3)^2 + \xi(x_3).$$

Diagonalizability of both the Ricci operator and the shape operator of the leafs of the foliation is a very restrictive condition, as the following result shows.

Theorem 4.23.   *The following assertions are equivalent:*

1. $\mathcal{M}_f$ *is locally symmetric.*

2. $\mathcal{N}^{\perp}$ is a totally geodesic bundle-like foliation.

3. $\mathcal{M}_f$ has diagonalizable Ricci tensor and $\mathcal{N}^{\perp}$ is totally geodesic.

Remark 4.24.   Locally symmetric Walker manifolds of non-zero scalar curvature are locally given by Theorem 4.6, which shows that the corresponding Ricci operator is diagonalizable. Hence, since the Ricci operator is parallel, the manifold splits locally into a product of an interval and a Lorentzian surface of constant sectional curvature. The converse is trivial.

4.4.2   CONTACT WALKER MANIFOLDS

If $\left(\frac{f_{12}}{\tau}\right)_1 \neq 0$, or equivalently if $\operatorname{div} N \neq 0$, then the 1 form $\omega = dx_2 - \frac{f_{12}}{\tau} dx_3$ defines a contact structure. Since $\mathcal{M}_f$ is Lorentzian, the contact structure ω has an associated para-contact structure (φ, ξ, η), i.e.,

$$\varphi^2 = \operatorname{Id} - \eta \otimes \xi, \qquad \eta(\xi) = 1$$

and the metric satisfies

$$g(\varphi X, \varphi Y) = -g(X, Y) + \eta(X)\eta(Y)$$

for all vector fields X, Y on M. Moreover, $g(\varphi X, Y) = d\omega(X, Y)$.

Remark 4.25.   Observe that N does not correspond to the Reeb vector field ξ of the para-contact structure. Let $i_\star$ denote interior multiplication. The Reeb vector field is characterized by $i_\xi d\omega = 0$ and $\eta(\xi) = 1$. Further, observe that

$$i_N d\omega = -\left(\frac{f_{12}}{\tau}\right)_2 dx_1 + \frac{f_{12}}{\tau}\left(\frac{f_{12}}{\tau}\right)_2 dx_2 + \frac{f_{12}}{\tau}\left(\frac{f_{12}}{\tau}\right)_1 dx_3$$

from where it follows that $i_N d\omega = 0$ if and only if $d\omega$ vanishes identically.

Now, proceeding as in [219], one has the following

Theorem 4.26.   *Let $(M, \varphi, \xi, \eta, g)$ be an almost para-contact manifold. The induced almost para-complex structure on $M \times \mathbb{R}$*

$$J\left(X, a\frac{d}{dr}\right) = \left(\varphi X + a\xi, \eta(X)\frac{d}{dr}\right)$$

becomes almost para-Hermitian with respect to the conformal metric $h^0 = e^{2r}h$, where h is the product metric

$$h\left[\left(X, a\frac{d}{dr}\right), \left(Y, b\frac{d}{dr}\right)\right] = g(X, Y) - ab.$$

Moreover, $(M \times \mathbb{R}, h^0, J)$ is almost para-Kaehler if and only if $(M, \varphi, \xi, \eta, g)$ is para-contact.

Remark 4.27. An alternative description of the conformal metric $h^0 = e^{2r}h$ in $M \times \mathbb{R}$, where $h = g_M - dr^2$ is the product metric, goes as follows. Put $v = e^r$. Then h^0 is just the warped product metric $-dv^2 + v^2 g_M$ and hence $(M \times \mathbb{R}, h^0)$ can be interpreted as the *cone* $\mathbb{R} \times_v M$ over the manifold (M, g_M). A special case of this result (the correspondence between para-Sasakian structures $(M, \varphi, \xi, \eta, g)$ and para-Kaehler cones) was discussed in [10].

4.5   STRICT WALKER MANIFOLDS

The results in Section 3.3 show that a 3 dimensional Walker manifold is a *strict Walker manifold* if and only if $\mathcal{D}$ admits a null parallel spanning vector field or, equivalently (see Remark 4.3), if we can choose a coordinate system so $f(x_1, x_2, x_3) = f(x_2, x_3)$ [255]. We assume henceforth that $\mathcal{M}_f$ is a strict Walker manifold and that (x_1, x_2, x_3) are adapted coordinates so $f = f(x_2, x_3)$. We may then use Lemma 4.2 to see

$$\nabla_{\partial_{x_2}} \partial_{x_3} = \tfrac{1}{2} f_2 \partial_{x_1} \quad \text{and} \quad \nabla_{\partial_{x_3}} \partial_{x_3} = \tfrac{1}{2} f_3 \partial_{x_1} - \tfrac{\varepsilon}{2} f_2 \partial_{x_2}.$$

Furthermore, since f does not depend on x_1, the relations of Lemma 4.4 simplify to become:

$$\mathcal{R}(\partial_{x_2}, \partial_{x_3}) \partial_{x_2} = \tfrac{1}{2} f_{22} \partial_{x_1} \quad \text{and} \quad \mathcal{R}(\partial_{x_2}, \partial_{x_3}) \partial_{x_3} = -\tfrac{\varepsilon}{2} f_{22} \partial_{x_2}.$$

By Lemma 4.4, the Ricci operator becomes

$$\rho = \begin{pmatrix} 0 & 0 & -\tfrac{\varepsilon}{2} f_{22} \\ 0 & 0 & 0 \\ 0 & 0 & 0 \end{pmatrix}.$$

Theorem 4.28. *Let $\mathcal{M}_f$ for $f = f(x_2, x_3)$ be a strict Walker manifold. Then:*

1. *All the scalar invariants of $\mathcal{M}_f$ vanish.*

2. *$\mathcal{M}_f$ is locally symmetric if and only if $f(x_2, x_3) = x_2^2 \alpha + x_2 \beta(x_3) + \xi(x_3)$ for α constant.*

3. *The following assertions are equivalent:*

   (a) *The Ricci tensor is Codazzi.*

   (b) *$\mathcal{M}_f$ is locally conformally flat.*

   (c) *We can choose a coordinate system so that $f(x_2, x_3) = x_2^2 \alpha(x_3) + x_2 \beta(x_3) + \xi(x_3)$.*

Proof. The covariant derivative of the curvature tensor is given by

$$(\nabla_{\partial_{x_2}}\mathcal{R})(\partial_{x_2}, \partial_{x_3})\partial_{x_2} = \tfrac{1}{2}f_{222}\partial_{x_1} \quad \text{and} \quad (\nabla_{\partial_{x_3}}\mathcal{R})(\partial_{x_2}, \partial_{x_3})\partial_{x_2} = \tfrac{1}{2}f_{223}\partial_{x_1}.$$

Hence, since ∂_{x_1} is parallel, it follows that the non-zero components of the higher order covariant derivatives of the curvature tensor produce higher order derivatives of f in the direction of ∂_{x_1}. Moreover, since the inverse of the Walker manifold satisfies

$$g_f^{-1} = \begin{pmatrix} -f(x_2, x_3) & 0 & 1 \\ 0 & \varepsilon & 0 \\ 1 & 0 & 0 \end{pmatrix} \tag{4.3}$$

it follows that $(g_f^{-1})(\partial_{x_2}, \partial_{x_3}) = 0$ so $\mathcal{M}_f$ is VSI as described in Section 2.4.3. This proves Assertion (1); the other assertions follow similarly. □

4.6   THREE DIMENSIONAL HOMOGENEOUS LORENTZIAN MANIFOLDS

Suppose that $\mathcal{M}$ is a 3 dimensional Riemannian symmetric space. Since the Ricci operator is parallel, either $\mathcal{M}$ has constant sectional curvature or $\mathcal{M}$ is locally isometric to a product of a real interval and a surface of constant scalar curvature. This result only generalizes to the Lorentzian setting in the special case when the Ricci operator is assumed to be diagonalizable. Let $\mathbb{S}^m$ and $\mathbb{H}^m$ be the Lorentzian space forms of dimension m and of constant sectional curvature $+1$ and -1, respectively. Walker geometry enters once again in the following result [67]:

Theorem 4.29. *Let $\mathcal{M}$ be a simply connected 3 dimensional Lorentzian symmetric space. Then $\mathcal{M}$ is one of the following:*

1. *A Lorentzian space form $\mathbb{S}^3$, $\mathbb{R}^3$ or $\mathbb{H}^3$.*

2. *A direct product $\mathbb{R} \times \mathbb{S}^2$, $\mathbb{R} \times \mathbb{H}^2$, $\mathbb{S}^2 \times \mathbb{R}$ or $\mathbb{H}^2 \times \mathbb{R}$.*

3. *A locally symmetric strict Walker manifold of dimension 3 as described in* Section 4.5.

Recall from Section 3.4.3 that a pseudo-Riemannian *homogeneous structure* on $\mathcal{M}$ is a tensor field T of type $(1, 2)$ such that the connection $\bar{\nabla} = \nabla - T$ satisfies

$$\bar{\nabla}g = 0, \quad \bar{\nabla}R = 0, \quad \bar{\nabla}T = 0. \tag{4.4}$$

The following result explains the geometric meaning of such structures [122].

Theorem 4.30. *Let $\mathcal{M}$ be a simply connected complete pseudo-Riemannian manifold. The following assertions are equivalent:*

  1. $\mathcal{M}$ *admits a homogeneous pseudo-Riemannian structure.*

  2. $\mathcal{M}$ *is a reductive homogeneous pseudo-Riemannian manifold $(G/H, g)$.*

  3. *The restriction of the Cartan-Killing form of $\mathfrak{so}(T_P M)$ to the Lie subalgebra $\mu(\mathfrak{h})$ is non-degenerate where $\mathfrak{h}$ denotes the Lie algebra of the isotropy group H and where μ is an appropriately chosen representation of $\mathfrak{h}$.*

Recall that $\mathcal{M}$ is said to be *locally symmetric* if $\nabla R = 0$. Locally symmetric pseudo-Riemannian manifolds are naturally equipped with the trivial $(T = 0)$ homogeneous structure. Lie groups are also characterized by possessing a special kind of homogeneous structure as follows [122].

Lemma 4.31.   *Let $\mathcal{M}$ be a simply connected complete pseudo-Riemannian manifold. If $\mathcal{M}$ admits a pseudo-Riemannian homogeneous structure T such that $T_X Y = \nabla_X Y$ for all X, Y vector fields tangent to $\mathcal{M}$, then $\mathcal{M}$ has a Lie group structure, unique up to isomorphism, and g is left invariant.*

Calvaruso [67] used this result to obtain the following description of complete simply connected homogeneous Lorentzian manifolds of dimension 3.

Theorem 4.32.   *Let $\mathcal{M}$ be a 3 dimensional simply connected complete homogeneous Lorentzian manifold. Then, either $\mathcal{M}$ is symmetric, or $\mathcal{M}$ is isometric to a 3 dimensional Lie group equipped with a left invariant Lorentzian metric.*

Combining Theorem 4.32 with results of [87, 227] leads to the classification of 3 dimensional homogeneous Lorentzian manifolds. Moreover 3 dimensional Lorentzian symmetric spaces are now classified by using previous results on Walker manifolds (cf. Theorem 4.6).

4.6.1   THREE DIMENSIONAL LIE GROUPS AND LIE ALGEBRAS

We recall the material of Section 2.3.1. In order to study 3 dimensional Lie algebras we will make use of the familiar *cross product operation*. If u and v are elements of a 3 dimensional vector space which is equipped with a positive define metric and with a preferred orientation, then the cross product $u \times v$ is defined. This product is bilinear and skew symmetric as a function of u and v. The vector $u \times v$ is orthogonal to both u and v and has length equal to the square root of the determinant $\langle u, u \rangle \langle v, v \rangle - \langle u, v \rangle^2$. Its direction is determined by the requirement that the triple $\{u, v, u \times v\}$ is positively oriented whenever u and v are linearly independent. The cross product in $\mathbb{R}^3$ corresponds to the product of imaginary quaternions $i \times j = k, j \times k = i, k \times i = j$ where

$$i^2 = -1, \quad j^2 = -1, \quad k^2 = -1, \quad ij = -ji = k.$$

Let $\{i, j, k\}$ be an orthonormal basis of signature $(1, 2)$ for Minkowski space $\mathbb{R}^3_1$. We then have $i \times j = -k, j \times k = i$, and $k \times i = j$. This is related to the algebra

$$i^2 = 1, \quad j^2 = 1, \quad k^2 = -1, \quad ij = -ji = k.$$

We have the following [201, 227]:

Lemma 4.33.   *Let G be a connected 3 dimensional Lie group with a positive definite left invariant metric. Choose an orientation for the Lie algebra of G, so that the cross product is defined. The bracket product operation in the Lie algebra $\mathfrak{g}$ of G satisfies $[u, v] = L(u \times v)$ where L is a uniquely defined linear mapping from $\mathfrak{g}$ to itself; G is unimodular if and only if L is self-adjoint.*

We now specialize to the unimodular case. If L is self-adjoint and if G is Riemannian, then there exists an orthonormal basis $\{e_1, e_2, e_3\}$ consisting of eigenvectors, $Le_i = \lambda_i e_i$. Replacing e_1 by $-e_1$ if necessary, we may assume that the basis $\{e_1, e_2, e_3\}$ is positively oriented. The bracket product operation is then given by $[e_1, e_2] = L(e_1 \times e_2) = \lambda_3 e_3$, with similar expression for other $[e_i, e_j]$. Thus we obtain the following normal form,

$$[e_2, e_3] = \lambda_1 e_1, \qquad [e_3, e_1] = \lambda_2 e_2, \qquad [e_1, e_2] = \lambda_3 e_3,$$

for the bracket product operation in a 3 dimensional unimodular Lie algebra with a positive definite metric. On the other hand, we emphasize that if the metric is Lorentzian, even if L is self-adjoint, L may have any one of the four Jordan normal forms already discussed for the Ricci operator in Section 4.3.2.

In the non-unimodular case, we have [87, 201]:

Lemma 4.34.   *Let G be a connected 3 dimensional Lie group which is not unimodular.*

1. *If G is Riemannian, there is a basis $\{e_1, e_2, e_3\}$ for $\mathfrak{g}$ with structure constants $\{\alpha, \beta, \gamma, \delta\}$ with $\alpha + \delta = 2$ so*

$$[e_1, e_2] = \alpha e_2 + \beta e_3, \quad [e_1, e_3] = \gamma e_2 + \delta e_3, \quad [e_2, e_3] = 0.$$

   If we exclude the exceptional case $\alpha = \delta = 1$ and $\beta = \gamma = 0$, then $\alpha\delta - \beta\gamma$ provides a complete isomorphism invariant for this Lie algebra.

2. *If G is Lorentzian, there is a basis $\{e_1, e_2, e_3\}$ for $\mathfrak{g}$ with structure constants $\{\alpha, \beta, \gamma, \delta\}$ with $\alpha + \delta \neq 0$ so*

$$[e_1, e_2] = 0, \quad [e_1, e_3] = \alpha e_1 + \beta e_2, \quad [e_2, e_3] = \gamma e_1 + \delta e_2.$$

Furthermore, one of the following holds:

   (a) *$\{e_1, e_2, e_3\}$ is orthonormal with $\langle e_1, e_1 \rangle = -\langle e_2, e_2 \rangle = -\langle e_3, e_3 \rangle = -1$ and the structure constants satisfy $\alpha\gamma - \beta\delta = 0$.*

   (b) *$\{e_1, e_2, e_3\}$ is orthonormal with $\langle e_1, e_1 \rangle = \langle e_2, e_2 \rangle = -\langle e_3, e_3 \rangle = 1$ and the structure constants satisfy $\alpha\gamma + \beta\delta = 0$.*

(c) $\{e_1, e_2, e_3\}$ is an orthonormal basis, $\alpha\gamma = 0$, and

$$\langle \cdot, \cdot \rangle = \begin{pmatrix} 1 & 0 & 0 \\ 0 & 0 & -1 \\ 0 & -1 & 0 \end{pmatrix}.$$

4.7   CURVATURE HOMOGENEOUS LORENTZIAN MANIFOLDS

Work of [11, 224] shows there exists a universal integer $k_{p,q}$ so that any pseudo-Riemannian manifold of signature (p, q) which is $k_{p,q}$ curvature homogeneous is necessarily locally homogeneous. In the 2 dimensional setting, 0 curvature homogeneity implies local homogeneity regardless of the signature. However, in dimension 3, the Riemannian and the Lorentzian settings differ. In the Riemannian context, Sekigawa [232] showed that 1 curvature homogeneity implies local homogeneity in dimension 3. On the other hand, Bueken and Vanhecke [65] constructed 1 curvature homogeneous Lorentzian manifolds which are not locally homogeneous. Bueken and Djorić [64] showed this result is sharp by showing that a 2 curvature homogeneous Lorentzian manifold of dimension 3 is necessarily locally homogeneous.

The main difference between the Riemannian and Lorentzian cases arises from the different behavior of the Ricci operator ρ. While ρ is always diagonalizable in the Riemannian setting, in the Lorentzian case the Ricci operator may take any of the four Jordan normal forms (Ia)–(III) discussed in Section 4.3.2. It is worth emphasizing here that all the algebraic possibilities for the Ricci operator can be realized geometrically by a non-homogeneous curvature homogeneous Lorentzian metric of dimension 3 [72] (except trivially the case (Ia) when the Ricci operator is a multiple of the identity).

The full classification of Lorentzian manifolds of dimension 3 which are curvature homogeneous up to order one was obtained by Bueken and Djorić in [64]. They proved that there exist exactly two classes of non-homogeneous 1 curvature homogeneous Lorentzian manifolds of dimension 3 whose Ricci operator has Jordan normal form corresponding to Type Ia and Type II, thus showing that not all the algebraic possibilities may be realized at the differentiable level in this situation. We report here their description.

4.7.1   DIAGONALIZABLE RICCI OPERATOR

Non-homogeneous 1 curvature homogeneous Lorentzian metrics in dimension 3 with diagonalizable Ricci operator have exactly two distinct Ricci curvatures. They are given by the following construction [64, 74].

Example 4.35. Let $\{e_1, e_2, e_3\}$ be a local orthonormal frame for TM where e_3 is timelike. Let G be a constant and let ψ be an arbitrary function. The metric is then described by the relations

$$[e_1, e_2] = -e_2 - (G+2)e_3, \quad [e_1, e_3] = -Ge_2 + e_3,$$
$$[e_2, e_3] = 2(G+1)e_1 - \psi e_2 - \psi e_3, \quad \text{where} \quad e_1(\psi) = (G+1)\psi.$$

We have [64, 74]:

Proposition 4.36. *Let M be a 3 dimensional Lorentzian manifold with ρ diagonalizable.*

1. *M is 1 curvature homogeneous but not homogeneous if and only if M is locally as in* Example 4.35.

2. *Let M be as in* Example 4.35.

   (a) *M is locally homogeneous if and only if ψ is constant.*

   (b) *M is never locally symmetric.*

   (c) *M admits a parallel degenerate line field if and only if $G = -1$.*

A pseudo-Riemannian manifold M is said to be *semi-symmetric* if its curvature tensor $\mathcal{R}$ satisfies $\mathcal{R}(X, Y) \cdot \mathcal{R} = 0$, for all vector fields X, Y on M. Here, $\mathcal{R}(X, Y)$ acts as a derivation on $\mathcal{R}$. The curvature tensor of a semi-symmetric manifold at any point is the same as the curvature tensor of a symmetric space (which may change with the point). A pseudo-Riemannian manifold M is said to be *Ricci semi-symmetric* if $\mathcal{R}(X, Y) \cdot \rho = 0$, for all vector fields X, Y on M, where ρ denotes the Ricci operator. Clearly, any semi-symmetric space is Ricci semi-symmetric and so, the class of Ricci semi-symmetric manifolds includes the one of semi-symmetric manifolds. The converse does not hold in general, but it is true in dimension 3, where the curvature tensor is completely determined by the Ricci tensor. Note that Ricci semi-symmetry is just the property of curvature-Ricci commuting (cf. Section 1.4.5). By Proposition 4.36, the Lorentzian metric of Example 4.35 with $G = -1$ has Ricci operator $\rho = \text{diag}[0, b, b]$ where $b = -(e_2 + e_3)\psi$, and thus it is curvature–curvature and curvature–Ricci commuting (cf. Theorem 4.16).

4.7.2   TYPE II RICCI OPERATOR

A non-homogeneous Lorentzian manifold in dimension 3 which is 1 curvature homogeneous and has non-diagonalizable Ricci operator must have exactly one single Ricci curvature which is a double root of the minimal polynomial of ρ [64, 74]. We recall the construction as follows.

Example 4.37. Let $\{e_1, e_2, e_3\}$ be a local orthonormal basis, with e_3 timelike, let θ be a function, let $\eta = \pm 1$, and let C and D be two constants such that

$$[e_1, e_2] = -(\theta + D)e_2 + \eta(C - \theta)e_3, \quad [e_1, e_3] = \eta(C + \theta)e_2 + (\theta - D)e_3, \quad [e_2, e_3] = 0.$$

Here $e_1(\theta) = \eta - 2(C + D)\theta$, and $(e_2 + \eta e_3)(\theta) = 0$.

The Ricci operator of any metric given by Example 4.37 has a Jordan normal form corresponding to Type II with a single eigenvalue $\lambda = -2D^2$. We have by [64, 74]:

Proposition 4.38. *Let $\mathcal{M}$ be a 3 dimensional Lorentzian manifold with ρ of Type II.*

1. *$\mathcal{M}$ is 1 curvature homogeneous but not homogeneous if and only if $\mathcal{M}$ locally is as in* Example 4.37.

2. *Let $\mathcal{M}$ be as in* Example 4.37.

   (a) *$\mathcal{M}$ is locally homogeneous if and only if either θ is constant, or $C = D = 0$ and θ satisfies $e_1(\theta) = \eta$ and $(e_2 + \eta e_3)(\theta) = 0$.*

   (b) *$\mathcal{M}$ admits a parallel degenerate line field if and only if $D = 0$. Moreover, in such a case $\mathcal{M}$ is a strict Walker manifold.*

   (c) *$\mathcal{M}$ is never locally symmetric.*

We may conclude that any Walker manifold as given by Example 4.37 with $D = 0$ has a 2-step nilpotent Ricci operator, and thus it is curvature–curvature commuting and curvature–Ricci commuting (Ricci semi-symmetric) as a consequence of Theorem 4.16. Finally, observe that any Lorentzian manifold $\mathcal{M}$ which is given by Example 4.37 is curvature–Ricci commuting (semi-symmetric) if and only if $D = 0$, and thus it is a Walker manifold (cf. [73]).

Remark 4.39. The existence of certain homogeneous structures influences the curvature. The converse is also true. For example, a 3 dimensional manifold is locally symmetric if and only if the Ricci tensor is parallel.

A close family of homogeneous structures is that of *naturally reductive* homogeneous structures (i.e., $T_X X = 0$ for all vector fields X). Moreover, in the Riemannian case any 3 dimensional manifold whose Ricci tensor is cyclic parallel is locally isometric to a naturally reductive homogeneous space [222]. Recent examples of Calvaruso show that the above result is not true in the Lorentzian setting. Indeed, a Lorentzian manifold $\mathcal{M}$ which has the form given by Example 4.37 has cyclic parallel Ricci tensor if and only if $C = D$ while metrics given in Example 4.35 never have cyclic parallel Ricci tensor [74].

CHAPTER 5

Four-Dimensional Walker Manifolds

5.1 INTRODUCTION

In this Chapter, we shall discuss Walker geometry in dimension 4. We present a brief review of some of the history of this subject in Section 5.2. Basic facts are presented in Section 5.3 where we discuss the Levi-Civita connection, geodesic equations, the curvature tensor, the Ricci tensor, the Einstein equations, self-duality, and anti-self-duality.

Section 5.4 treats para-Hermitian structures. We have seen in Chapter 3 that para-Kaehler and hyper-symplectic metrics are necessarily of Walker type. This motivates the consideration of Walker metrics in connection with almost para-Hermitian structures. It is shown that any 4 dimensional Walker manifold of nowhere zero scalar curvature has a natural almost para-Hermitian structure. In contrast to the Goldberg-Sachs theorem, if this structure is both self-dual and $\star$-Einstein, it is symplectic but not necessarily integrable. This is due to the non-diagonalizability of the self-dual Weyl conformal curvature tensor. The isotropic condition is introduced, Chern classes are discussed, and self-duality is treated in relation to para-Hermitian geometry.

It is one of the features of Walker geometry in signature $(2, 2)$ that there is a canonical orientation. Let $\mathcal{M}$ be a pseudo-Riemannian manifold of signature $(2, 2)$ which admits a null parallel distribution $\mathcal{D}$. Let $\mathcal{E}$ be any complementary distribution so

$$TM = \mathcal{D} \oplus \mathcal{E} \ .$$

If $\{e_1, e_2\}$ is any basis for $\mathcal{D}$, we may specify a basis $\{f_1, f_2\}$ for $\mathcal{E}$ by requiring that $g(e_i, f_j) = \delta_{ij}$. The orientation corresponding to

$$e_1 \wedge e_2 \wedge f_1 \wedge f_2$$

is then independent of the choice $\{e_1, e_2\}$ and of the complementary distribution chosen. We adopt this canonical orientation henceforth; it will be crucial in what follows that the notions of self-dual and anti-self-dual are quite different and are not interchangeable.

5.2 HISTORY

The influence of the curvature on a manifold can be measured in many different ways and it appears in many different contexts. An example of such influence is the existence of additional structures on the manifold under certain curvature conditions. This is specially clear in dimension 4 when one describes the curvature tensor as an endomorphism of $\Lambda^2(M)$. Since $R : \Lambda^2 \to \Lambda^2$ is self-adjoint, the existence of distinguished eigenvalues gives rise to locally defined 2 forms and hence, to almost Hermitian structures. Properties of such structures can be specialized from curvature conditions like Einstein or locally conformally flat.

Derdzinski [97] showed that any oriented Einstein 4 dimensional manifold whose self-dual curvature operator is degenerate is naturally equipped with a Hermitian structure which becomes locally conformally Kaehler on some open dense set where $|W^+| \neq 0$. This result was later used by Apostolov and Gauduchon [14] and extended to the pseudo-Riemannian setting in [12] under some additional conditions on the diagonalizability of the self-dual Weyl curvature operator.

An additional remark in the pseudo-Riemannian case comes from the fact that the induced metric on $\Lambda^\pm(M)$ has Lorentzian signature, and hence the causal character of the distinguished eigenspace of W^+ must be considered. Spacelike 2 forms on $\Lambda^-(M)$ define almost Hermitian structures while timelike 2 forms on $\Lambda^+(M)$ define almost para-Hermitian structures. In both cases the generalized Goldberg-Sachs theorem holds under the diagonalizability assumption on the corresponding Weyl operators [12, 173]. Recently it has been shown in [91] that the diagonalizability of the self-dual Weyl operator cannot be removed by constructing explicit examples of self-dual Einstein 4 dimensional manifolds whose self-dual curvature operator is degenerate and the associated almost para-Hermitian structure is never integrable but symplectic. Walker metrics are the underlying structure behind such phenomena as discussed in Section 5.4.

We also refer to related work in [79, 108, 124].

5.3   FOUR-DIMENSIONAL WALKER MANIFOLDS

As we have seen in Section 4.3.1, the Walker coordinates introduced in Section 3.3 become simpler when the null r-plane $\mathcal{D}$ has maximum dimensionality. Since dim $\mathcal{D} = r \leq \frac{m}{2}$, two cases occur depending on whether m is odd or even. We have devoted Chapter 4 to the study of the case $m = 3$, we now turn our attention to the case $m = 4$. We begin with the following general result [255]:

Theorem 5.1.   *A canonical form for a $2n$ dimensional pseudo-Riemannian manifold $\mathcal{M}$ admitting a parallel field of null n dimensional planes $\mathcal{D}$ is given by the metric tensor:*

$$(g_{ij}) = \begin{pmatrix} 0 & \mathrm{Id}_n \\ \mathrm{Id}_n & B \end{pmatrix}$$

where Id_n is the $n \times n$ identity matrix and B is a symmetric $n \times n$ matrix whose entries are functions of the coordinates $(x_1, \dots, x_{2n})$.

We note that if the null n-plane is strictly parallel, then the entries of B in the previous theorem can be chosen to be independent of the coordinates $(x_1, \dots, x_n)$ [255] – we also refer to Theorem 3.2). We focus on the geometry of 4 dimensional pseudo-Riemannian manifolds admitting a parallel degenerate 2 dimensional plane field, i.e., the simplest even dimensional Walker manifold admitting a null parallel plane of maximum dimensionality. Due to Theorem 5.1, in this case the

Walker coordinates in Section 3.3 can be further specialized. We adopt the following notation for the remainder of this book.

Example 5.2.   Let $\mathcal{O}$ be an open subset of $\mathbb{R}^4$. Let $a, b, c \in C^\infty(\mathcal{O})$ be smooth functions on $\mathcal{O}$. We set $\mathcal{M}_{a,b,c} := (\mathcal{O}, g_{a,b,c})$ where

$$
\begin{aligned}
g_{a,b,c} \quad := \quad & 2(dx_1 \circ dx_3 + dx_2 \circ dx_4) + a(x_1, x_2, x_3, x_4)dx_3 \circ dx_3 \\
+ \quad & b(x_1, x_2, x_3, x_4)dx_4 \circ dx_4 + 2c(x_1, x_2, x_3, x_4)dx_3 \circ dx_4 \, .
\end{aligned}
$$

A straightforward calculation shows

Theorem 5.3.   *The Christoffel symbols of $\mathcal{M}_{a,b,c}$ are given by:*

$$\nabla_{\partial_{x_1}} \partial_{x_3} = \tfrac{1}{2}a_1 \partial_{x_1} + \tfrac{1}{2}c_1 \partial_{x_2}, \qquad \nabla_{\partial_{x_1}} \partial_{x_4} = \tfrac{1}{2}c_1 \partial_{x_1} + \tfrac{1}{2}b_1 \partial_{x_2},$$

$$\nabla_{\partial_{x_2}} \partial_{x_3} = \tfrac{1}{2}a_2 \partial_{x_1} + \tfrac{1}{2}c_2 \partial_{x_2}, \qquad \nabla_{\partial_{x_2}} \partial_{x_4} = \tfrac{1}{2}c_2 \partial_{x_1} + \tfrac{1}{2}b_2 \partial_{x_2},$$

$$\nabla_{\partial_{x_3}} \partial_{x_3} = \tfrac{1}{2}(aa_1 + ca_2 + a_3)\partial_{x_1} + \tfrac{1}{2}(ca_1 + ba_2 - a_4 + 2c_3)\partial_{x_2} - \tfrac{a_1}{2}\partial_{x_3} - \tfrac{a_2}{2}\partial_{x_4},$$

$$\nabla_{\partial_{x_3}} \partial_{x_4} = \tfrac{1}{2}(a_4 + ac_1 + cc_2)\partial_{x_1} + \tfrac{1}{2}(b_3 + cc_1 + bc_2)\partial_{x_2} - \tfrac{c_1}{2}\partial_{x_3} - \tfrac{c_2}{2}\partial_{x_4},$$

$$\nabla_{\partial_{x_4}} \partial_{x_4} = \tfrac{1}{2}(ab_1 + cb_2 - b_3 + 2c_4)\partial_{x_1} + \tfrac{1}{2}(cb_1 + bb_2 + b_4)\partial_{x_2} - \tfrac{b_1}{2}\partial_{x_3} - \tfrac{b_2}{2}\partial_{x_4}.$$

The following is now immediate:

Lemma 5.4.   *A curve $\gamma(t) = (x_1(t), x_2(t), x_3(t), x_4(t))$ in $\mathcal{M}_{a,b,c}$ is a geodesic if and only if the following equations are satisfied:*

$$0 = \ddot{x}_1 + \dot{x}_1\dot{x}_3 a_1 + \dot{x}_1\dot{x}_4 c_1 + \dot{x}_2\dot{x}_3 a_2 + \dot{x}_2\dot{x}_4 c_2 + \tfrac{1}{2}\dot{x}_3\dot{x}_3(a_3 + ca_2 + aa_1)$$

$$+\dot{x}_3\dot{x}_4(a_4 + cc_2 + ac_1) + \tfrac{1}{2}\dot{x}_4\dot{x}_4(2c_4 - b_3 + cb_2 + ab_1),$$

$$0 = \ddot{x}_2 + \dot{x}_1\dot{x}_3 c_1 + \dot{x}_1\dot{x}_4 b_1 + \dot{x}_2\dot{x}_3 c_2 + \dot{x}_2\dot{x}_4 b_2 + \tfrac{1}{2}\dot{x}_3\dot{x}_3(2c_3 - a_4 + ba_2 + ca_1)$$

$$+\dot{x}_3\dot{x}_4(b_3 + bc_2 + cc_1) + \tfrac{1}{2}\dot{x}_4\dot{x}_4(b_4 + bb_2 + cb_1),$$

$$0 = \ddot{x}_3 - \tfrac{1}{2}\dot{x}_3\dot{x}_3 a_1 - \dot{x}_3\dot{x}_4 c_1 - \tfrac{1}{2}\dot{x}_4\dot{x}_4 b_1, \qquad 0 = \ddot{x}_4 - \tfrac{1}{2}\dot{x}_3\dot{x}_3 a_2 - \dot{x}_3\dot{x}_4 c_2 - \tfrac{1}{2}\dot{x}_4\dot{x}_4 b_2.$$

Henceforth, we shall only give the (possibly) non-zero components of various tensors up to the usual $\mathbb{Z}_2$ symmetries. We have

Theorem 5.5.

1. *The curvature tensor of $\mathcal{M}_{a,b,c}$ is given by:*

$$R_{1313} = \tfrac{1}{2}a_{11}, \quad R_{1314} = \tfrac{1}{2}c_{11}, \quad R_{1323} = \tfrac{1}{2}a_{12}, \quad R_{1324} = \tfrac{1}{2}c_{12}, \qquad R_{1424} = \tfrac{1}{2}b_{12},$$

$$R_{1334} = -\tfrac{1}{4}\left(-a_2 b_1 + c_1 c_2 + 2a_{14} - 2c_{13}\right), \quad R_{1414} = \tfrac{1}{2}b_{11}, \quad R_{1423} = \tfrac{1}{2}c_{12},$$

$$R_{1434} = -\tfrac{1}{4}\left(-c_1^2 + a_1 b_1 - b_1 c_2 + b_2 c_1 - 2b_{13} + 2c_{14}\right), \quad R_{2323} = \tfrac{1}{2}a_{22},$$

$$R_{2334} = -\tfrac{1}{4}\left(c_2^2 - a_2 b_2 - a_1 c_2 + a_2 c_1 + 2a_{24} - 2c_{23}\right), \quad R_{2424} = \tfrac{1}{2}b_{22},$$

$$R_{2324} = \tfrac{1}{2}c_{22},$$

$$R_{2434} = -\tfrac{1}{4}\left(a_2 b_1 - c_1 c_2 - 2b_{23} + 2c_{24}\right),$$

$$R_{3434} = -\tfrac{1}{4}\left(-a c_1^2 - b c_2^2 + a a_1 b_1 + c a_1 b_2 - a_1 b_3 + 2a_1 c_4 + c a_2 b_1 + b a_2 b_2 + a_2 b_4\right.$$

$$\left.+a_3 b_1 - a_4 b_2 - 2a_4 c_1 + 2b_2 c_3 - 2b_3 c_2 - 2c c_1 c_2 - 2a_{44} - 2b_{33} + 4c_{34}\right).$$

2. *The Ricci tensor and the scalar curvature of* $\mathcal{M}_{a,b,c}$ *are given by:*

$$\rho_{13} = \tfrac{1}{2}(a_{11} + c_{12}), \quad \rho_{14} = \tfrac{1}{2}(b_{12} + c_{11}), \quad \rho_{23} = \tfrac{1}{2}(a_{12} + c_{22}),$$

$$\rho_{24} = \tfrac{1}{2}(b_{22} + c_{12}),$$

$$\rho_{33} = \tfrac{1}{2}(-c_2^2 + a_1 c_2 + a_2 b_2 - a_2 c_1 + a a_{11} + 2c a_{12} + b a_{22} + 2c_{23} - 2a_{24}),$$

$$\rho_{34} = \tfrac{1}{2}(-a_2 b_1 + c_1 c_2 + a_{14} + b_{23} + a c_{11} + 2c c_{12} - c_{13} + b c_{22} - c_{24}),$$

$$\rho_{44} = \tfrac{1}{2}(-c_1^2 + a_1 b_1 - b_1 c_2 + b_2 c_1 + a b_{11} + 2c b_{12} - 2b_{13} + b b_{22} + 2c_{14}),$$

$$\tau = a_{11} + b_{22} + 2c_{12}.$$

3. *Let* $\rho_0 := \rho - \tfrac{\tau}{4}g_{a,b,c}$ *be the trace free Ricci tensor of* $\mathcal{M}_{a,b,c}$.

$$(\rho_0)_{13} = -(\rho_0)_{24} = (\rho_0)_{31} = -(\rho_0)_{42} = \tfrac{1}{4}\left(a_{11} - b_{22}\right),$$

$$(\rho_0)_{14} = (\rho_0)_{41} = \tfrac{1}{2}\left(b_{12} + c_{11}\right), \quad (\rho_0)_{23} = (\rho_0)_{32} = \tfrac{1}{2}\left(a_{12} + c_{22}\right),$$

$$(\rho_0)_{33} = \tfrac{1}{4}\left(2a_1 c_2 + 2a_2 b_2 - 2a_2 c_1 - 2c_2^2 + a a_{11}\right.$$

$$\left.+4c a_{12} + 2b a_{22} - 4a_{24} - a b_{22} - 2a c_{12} + 4c_{23}\right),$$

$$(\rho_0)_{34} = (\rho_0)_{43} = \tfrac{1}{4}\left(-2a_2 b_1 + 2c_1 c_2 - c a_{11} + 2a_{14} - c b_{22}\right.$$

$$\left.+2b_{23} + 2a c_{11} + 2c c_{12} - 2c_{13} + 2b c_{22} - 2c_{24}\right),$$

$$(\rho_0)_{44} = \tfrac{1}{4}\left(2a_1 b_1 - 2b_1 c_2 + 2b_2 c_1 - 2c_1^2 - b a_{11}\right.$$

$$\left.+2a b_{11} + 4c b_{12} - 4b_{13} + b b_{22} - 2b c_{12} + 4c_{14}\right).$$

4. $\mathcal{M}_{a,b,c}$ *is Einstein if and only if*

$$a_{11} - b_{22} = 0, \quad b_{12} + c_{11} = 0, \quad a_{12} + c_{22} = 0,$$

$$a_1 c_2 + a_2 b_2 - a_2 c_1 - c_2^2 + 2ca_{12} + ba_{22} - 2a_{24} - ac_{12} + 2c_{23} = 0,$$

$$a_2 b_1 - c_1 c_2 + ca_{11} - a_{14} - b_{23} - ac_{11} - cc_{12} + c_{13} - bc_{22} + c_{24} = 0,$$

$$a_1 b_1 - b_1 c_2 + b_2 c_1 - c_1^2 + ab_{11} + 2cb_{12} - 2b_{13} - bc_{12} + 2c_{14} = 0.$$

The PDE system in Theorem 5.5 (4) is hard to handle and only partial solutions are known for some special classes of Walker manifolds as will be considered in Chapters 7 and 8. Let W denote the Weyl conformal curvature tensor of Equation (1.10). An orthonormal basis can be specialized by using the canonical coordinates as follows:

$$e_1 := \tfrac{1}{2}(1 - a)\partial_{x_1} + \partial_{x_3}, \qquad e_2 := -c\partial_{x_1} + \tfrac{1}{2}(1 - b)\partial_{x_2} + \partial_{x_4},$$
$$e_3 := -\tfrac{1}{2}(1 + a)\partial_{x_1} + \partial_{x_3}, \qquad e_4 := -c\partial_{x_1} - \tfrac{1}{2}(1 + b)\partial_{x_2} + \partial_{x_4}.$$

(5.1)

Let $e_1 \wedge e_2 \wedge e_3 \wedge e_4$ orient $\mathcal{M}$ and let $\{E_i^{\pm}\}$ be as in Equation (1.12); this gives the canonical orientation of a Walker manifold of signature (4, 4). We adopt the notation of Equation (1.13).

Theorem 5.6. *The components of W^- for $\mathcal{M}_{a,b,c}$ are given by*

$$W_{11}^- = \tfrac{1}{12}(a_{11} + 3a_{22} + 3b_{11} + b_{22} - 4c_{12}), \qquad W_{22}^- = \tfrac{1}{6}(a_{11} + b_{22} - 4c_{12}),$$

$$W_{33}^- = -\tfrac{1}{12}(a_{11} - 3a_{22} - 3b_{11} + b_{22} - 4c_{12}), \qquad W_{12}^- = -\tfrac{1}{4}(a_{12} + b_{12} - c_{11} - c_{22}),$$

$$W_{13}^- = -\tfrac{1}{4}(a_{22} - b_{11}), \qquad W_{23}^- = \tfrac{1}{4}(a_{12} - b_{12} + c_{11} - c_{22}).$$

A complete description of self-dual Walker manifolds was obtained in [95, 106] after integrating the PDE system in Theorem 5.6. This yields:

Theorem 5.7. *$\mathcal{M}_{a,b,c}$ is self-dual if and only if*

$$a = x_1^3 \mathcal{A}(x_3, x_4) + x_1^2 \mathcal{B}(x_3, x_4) + x_1^2 x_2 \mathcal{C}(x_3, x_4) + x_1 x_2 \mathcal{D}(x_3, x_4)$$

$$+ x_1 P(x_3, x_4) + x_2 Q(x_3, x_4) + \xi(x_3, x_4),$$

$$b = x_2^3 \mathcal{C}(x_3, x_4) + x_2^2 \mathcal{E}(x_3, x_4) + x_1 x_2^2 \mathcal{A}(x_3, x_4) + x_1 x_2 \mathcal{F}(x_3, x_4)$$

$$+ x_1 S(x_3, x_4) + x_2 T(x_3, x_4) + \eta(x_3, x_4),$$

$$c = \tfrac{1}{2}x_1^2 \mathcal{F}(x_3, x_4) + \tfrac{1}{2}x_2^2 \mathcal{D}(x_3, x_4) + x_1^2 x_2 \mathcal{A}(x_3, x_4) + x_1 x_2^2 \mathcal{C}(x_3, x_4)$$

$$+\tfrac{1}{2}x_1x_2(\mathcal{B}(x_3,x_4)+\mathcal{E}(x_3,x_4))+x_1U(x_3,x_4)+x_2V(x_3,x_4)+\gamma(x_3,x_4).$$

We refer to Theorem 6.21 for a coordinate free description of self-dual Walker manifolds in terms of modified Riemannian extensions. Anti-self-dual Walker manifolds are much more difficult to describe. They have vanishing scalar curvature and their structure is, in some sense, richer than in the self-dual case. This uses the fact that the anti-self-dual Weyl curvature operator W^- allows complex eigenvalues (which may be constant for suitable Walker manifolds) [52]. We adopt the notation of Equation (1.13).

Theorem 5.8. *Let $M = M_{a,b,c}$.*

1. *The eigenvalues of W^+ are $\tau\left\{-\tfrac{1}{6},\tfrac{1}{12},\tfrac{1}{12}\right\}$.*

2. *The components of W^+ are given by:*

$$W_{11}^+ = -\tfrac{1}{12}(6ca_1b_2 - 6a_1b_3 - 6ba_1c_2 + 12a_1c_4 - 6ca_2b_1 + 6a_2b_4$$

$$+6ba_2c_1 + 6a_3b_1 - 6a_4b_2 - 12a_4c_1 + 6ab_1c_2 - 6ab_2c_1$$

$$+12b_2c_3 - 12b_3c_2 - a_{11} - 12c^2a_{11} - 12bca_{12} + 24ca_{14}$$

$$-3b^2a_{22} + 12ba_{24} - 12a_{44} - 3a^2b_{11} + 12ab_{13} - b_{22}$$

$$-12b_{33} + 12acc_{11} - 2c_{12} + 6abc_{12} - 24cc_{13} - 12ac_{14}$$

$$-12bc_{23} + 24c_{34}),$$

$$W_{12}^+ = W_{23}^+ = -\tfrac{1}{4}(-2ca_{11} - ba_{12} + 2a_{14} + ab_{12} - 2b_{23} + ac_{11}$$

$$-2cc_{12} - 2c_{13} - bc_{22} + 2c_{24}),$$

$$W_{22}^+ = \tfrac{\tau}{6}, \qquad W_{33}^+ = W_{11}^+ - \tfrac{\tau}{6}, \qquad W_{13}^+ = W_{11}^+ - \tfrac{\tau}{12}.$$

An explicit integration of the equations in Theorem 5.8 seems difficult. Partial results are however known for special families of Walker manifolds [111, 112]. By Theorem 5.8:

$$(W^+ + \tfrac{\tau}{6}\,\mathrm{Id})\cdot(W^+ - \tfrac{\tau}{12}\,\mathrm{Id}) = \tfrac{1}{48}(\tau^2 - 12\tau W_{11}^+ + 48(W_{12}^+)^2)\begin{pmatrix} -1 & 0 & -1 \\ 0 & 0 & 0 \\ 1 & 0 & 1 \end{pmatrix}.$$

Corollary 5.9. *Let $M = M_{a,b,c}$.*

1. *If $\tau \neq 0$, then*

(a) W^+ has non-zero eigenvalues $\left\{-\frac{\tau}{6}, \frac{\tau}{12}, \frac{\tau}{12}\right\}$.

(b) $\tau^2 - 12\tau W_{11}^+ + 48\left(W_{12}^+\right)^2 = 0$ if and only if W^+ is diagonalizable.

2. If $\tau = 0$, then W^+ vanishes if and only if $W_{11}^+ = W_{12}^+ = 0$.

Remark 5.10. Assertion (1) gives necessary and sufficient conditions for the diagonalizability of W^+. If $\tau = 0$, then W^+ is a 2-step nilpotent operator if and only if $W_{11}^+ \neq 0$ and $W_{12}^+ = 0$, while W^+ is a 3-step nilpotent operator if and only if $W_{12}^+ \neq 0$. We once again emphasize that signature $(2, 2)$ Walker manifolds inherit natural orientations and one can not interchange the notions of self-dual and of anti-self dual.

5.4 ALMOST PARA-HERMITIAN GEOMETRY

This section contains joint work with A. Cortés-Ayaso and J. C. Díaz-Ramos [91]. Let $\mathcal{M}_{a,b,c}$ be the Walker manifold of Example 5.2. The fact that para-Kaehler and hyper-symplectic metrics are necessarily of Walker type motivates the consideration of these metrics in connection with almost para-Hermitian structures. Moreover, as Theorem 5.8 shows, the self-dual Weyl curvature operator of any Walker manifold of non-zero scalar curvature has a distinguished eigenvalue. Metrics with such property have been broadly investigated in connection with the so-called Goldberg-Sachs Theorem [12, 14, 97, 173], which ensures the integrability of certain almost Hermitian or almost para-Hermitian structures under suitable curvature conditions. As we will show in this section, the diagonalizability of the self-dual curvature operator plays a fundamental role in such results.

After replacing the manifold $\mathcal{M}_{a,b,c}$ by a 2-fold covering if necessary, it follows that the $(-\frac{1}{6}\tau)$-eigenspace of W^+ is generated by a globally defined self-dual 2 form. We take as generator the 2 form

$$\Omega = -\frac{8W_{12}^+}{\tau\sqrt{2}}E_1^+ + \sqrt{2}E_2^+ + \frac{8W_{12}^+}{\tau\sqrt{2}}E_3^+;$$

$\langle \Omega, \Omega \rangle = -2$ with respect to the induced metric on Λ^+. Hence, it gives rise to an almost para-Hermitian structure $J \in \text{End}(TM)$ which is characterized by the identity:

$$g(X, JY) = \Omega(X, Y) \quad \text{for all} \quad X, Y .$$

The structure J has the following matrix representation with respect to the frame $\{e_1, e_2, e_3, e_4\}$ of Equation (5.1):

$$J = \begin{pmatrix} 0 & -\frac{4}{\tau}W_{12}^+ & -1 & -\frac{4}{\tau}W_{12}^+ \\ \frac{4}{\tau}W_{12}^+ & 0 & \frac{4}{\tau}W_{12}^+ & -1 \\ -1 & \frac{4}{\tau}W_{12}^+ & 0 & \frac{4}{\tau}W_{12}^+ \\ -\frac{4}{\tau}W_{12}^+ & -1 & -\frac{4}{\tau}W_{12}^+ & 0 \end{pmatrix}.$$

From here, it is easy to check that $J^2 = \mathrm{Id}$ and $g(JX, JY) = -g(X, Y)$ for all vectors X and Y. In what follows we will refer to J as *the almost para-Hermitian Walker structure* on $\mathcal{M}_{a,b,c}$; it is globally defined on any Walker manifold with nowhere zero scalar curvature. In Theorem 5.13, we discuss this structure in further detail.

Remark 5.11. A coordinate description of the almost para-Hermitian Walker structure and the associated 2 form is given by:

$$J = \begin{pmatrix} 1 & 0 & a & 2c - \frac{4}{\tau}W_{12}^+ \\ 0 & 1 & \frac{4}{\tau}W_{12}^+ & b \\ 0 & 0 & -1 & 0 \\ 0 & 0 & 0 & -1 \end{pmatrix}, \tag{5.2}$$

$$\Omega = -dx_1 \wedge dx_3 - dx_2 \wedge dx_4 + \left(c - \frac{4}{\tau}W_{12}^+\right) dx_3 \wedge dx_4.$$

5.4.1   ISOTROPIC ALMOST PARA-HERMITIAN STRUCTURES

We recall from Section 2.5.1 that (M, g, J) is *almost Kaehler* if $d\Omega = 0$. We say that a para-Hermitian structure (g, J) is an *isotropic para-Kaehler structure* if $|\nabla J|^2 = 0$ (equivalently, one has that $|\nabla\Omega|^2 = 0$) but $\nabla J \neq 0$ [131]. The existence of isotropic Kaehler structures was discussed previously [28, 93] in connection with the Goldberg Conjecture in the pseudo-Riemannian situation. This is closely related to the existence of Einstein and $\star$-Einstein almost Kaehler structures that are not Kaehler.

Example 5.12. Let $\tilde{\mathcal{C}} := (\mathcal{O}, g_{a,b,c}, J)$ where $(\mathcal{O}, g_{a,b,c})$ is as in Example 5.2 and where J is given by Equation (5.2), i.e.,

$$g_{a,b,c} := 2(dx_1 \circ dx_3 + dx_2 \circ dx_4) + a(x_1, x_2, x_3, x_4)dx_3 \circ dx_3$$
$$+ b(x_1, x_2, x_3, x_4)dx_4 \circ dx_4 + 2c(x_1, x_2, x_3, x_4)dx_3 \circ dx_4,$$

$$J = \begin{pmatrix} 1 & 0 & a & 2c - \frac{4}{\tau}W_{12}^+ \\ 0 & 1 & \frac{4}{\tau}W_{12}^+ & b \\ 0 & 0 & -1 & 0 \\ 0 & 0 & 0 & -1 \end{pmatrix}.$$

Theorem 5.13. $\tilde{\mathcal{C}}$ *is an isotropic para-Kaehler, isotropic almost para-Kaehler and isotropic para-Hermitian manifold, that is* $|\nabla\Omega|^2 = |d\Omega|^2 = |N_J|^2 = 0$.

Remark 5.14. We emphasize that although $\nabla\Omega$, $d\Omega$ and N_J are isotropic, these tensor fields need not vanish. If $f := \frac{4}{\tau}W_{12}^+$, then one has:

1. Ω is symplectic (equivalently, $\tilde{C}$ is *almost para-Kaehler*) if and only if $c + f$ does not depend on x_1 and x_2.

2. J is integrable (equivalently, $\tilde{C}$ is *para-Hermitian*) if and only if

   (a) $(2c_1 + f_1)a - ba_2 - 2ca_1 - (a_1 + 2c_2 + f_2)f + a_4 - 4c_3 - 2f_3 = 0.$

   (b) $ab_1 + bf_2 + 2cf_1 + (f_1 - b_2)f - 2b_3 - 2f_4 = 0.$

Moreover, $\tilde{C}$ is para-Kaehler if and only if the conditions in (1) and (2) above are satisfied.

5.4.2 CHARACTERISTIC CLASSES

Let M be a compact 4 dimensional manifold. Let $L[M]$ be the *Hirzebruch signature* of M and let $\chi[M]$ be the *Euler characteristic* of M. The Hitchin-Thorpe inequalities for Riemannian Einstein manifolds have been generalized [188] to yield the following inequality if $\mathcal{M}$ is a compact Einstein manifold of signature $(2,2)$:

$$\tfrac{3}{2}|L[M]| \leq \chi[M] \ .$$

Walker manifolds of nowhere zero scalar curvature admit almost anti-Hermitian structures, and hence the corresponding odd Chern numbers vanish [42, 45], thus providing examples where the equality is attained.

Theorem 5.15. *Let $\tilde{C}$ be as in Example 5.12. If τ is nowhere zero, then $\tfrac{3}{2}L[M] = -\chi[M]$.*

Osserman manifolds are Einstein, and thus the Euler characteristic $\chi[M]$ and the Hirzebruch signature $L[M]$ can be expressed as follows [187]:

$$
\begin{aligned}
\chi[M] &= -\frac{1}{8\pi^2} \int_M \left\{ \mathrm{Tr}[(W^+)^2] + \mathrm{Tr}[(W^-)^2] + \frac{\tau^2}{24} \right\}, \\
L[M] &= \frac{2}{3}\frac{1}{8\pi^2} \int_M \{ \mathrm{Tr}[(W^+)^2] - \mathrm{Tr}[(W^-)^2] \}.
\end{aligned}
\tag{5.3}
$$

Observe from Equation (5.3) that the Euler characteristic of any compact Einstein metric of signature $(2,2)$ is non-positive, provided that $W^\pm$ are not of Type Ib. Moreover, since Osserman metrics are self-dual or anti-self-dual, a detailed examination of Equation (5.3) using Theorem 5.15 shows the following result [49]:

Theorem 5.16. *Let $\mathcal{M}$ be a compact Jordan Osserman manifold of signature $(2,2)$. Then either $\mathcal{M}$ has constant sectional curvature or the Jacobi operator of $\mathcal{M}$ is either a 2 or a 3-step nilpotent operator.*

Further observe that the Jordan normal form of the Jacobi operator may change from point to point in an Osserman manifold. Hence, we have:

Theorem 5.17. *Let $\mathcal{M}$ be a compact Osserman manifold of signature $(2, 2)$. Then $L[M] = 0$ and the Jacobi operator has only one eigenvalue, which may be a single, double or triple root of the minimal polynomial. Moreover, $\chi[M] \le 0$, and $\chi[M] = 0$ if and only if the Jacobi operator is a nilpotent operator.*

5.4.3   SELF-DUAL WALKER MANIFOLDS

Let $\rho^\star$ and $\tau^\star$ be defined by Equation (1.9):

$$\rho_{ij}^\star := -g^{k\ell} R(e_k, e_i, Je_j, Je_\ell) \quad \text{and} \quad \tau^\star := g^{ij}\rho_{ij}^\star = \text{Tr}\{\rho^\star\} \, .$$

An almost para-Hermitian structure is said to be *weakly $\star$-Einstein* if $\rho^\star = \frac{1}{4}(\tau^\star)g$. Furthermore, if the $\star$-scalar curvature is constant, the structure is called *$\star$-Einstein*.

Theorem 5.18. *Assume that the manifold $\tilde{C}$ of Example 5.12 is self-dual so (a, b, c) have the form of Theorem 5.7. Then:*

1. *$\tilde{C}$ is weakly $\star$-Einstein if and only if*

$$a = x_1^2 \mathcal{B}(x_3, x_4) + x_1 P(x_3, x_4) + x_2 Q(x_3, x_4) + \xi(x_3, x_4),$$

$$b = x_2^2 \mathcal{B}(x_3, x_4) + x_1 S(x_3, x_4) + x_2 T(x_3, x_4) + \eta(x_3, x_4),$$

$$c = x_1 x_2 \mathcal{B}(x_3, x_4) + x_1 U(x_3, x_4) + x_2 V(x_3, x_4) + \gamma(x_3, x_4).$$

2. *$\tilde{C}$ is $\star$-Einstein if and only if $\mathcal{B}$ is constant.*

3. *If $\tilde{C}$ is $\star$-Einstein, then $\tilde{C}$ is almost para-Kaehler.*

Remark 5.19. Recall that the *para-holomorphic sectional curvature $K(X)$* which is associated with a non-null tangent vector field X is, by definition, the sectional curvature of the 2-plane given by setting $\pi := \text{Span}\{X, JX\}$, that is,

$$K(X) = K(\pi) := -g(\mathcal{R}(JX, X)X, JX)/g(X, X)^2 \, .$$

We use Theorem 5.18 to see that the para-holomorphic sectional curvature of any weakly $\star$-Einstein self-dual Walker manifold satisfies $K(X) = \mathcal{B}$, and thus it is pointwise constant, and globally constant if $\tilde{C}$ is further assumed to be $\star$-Einstein.

Remark 5.20.   Einstein self-dual Walker manifolds have been studied in [106] in connection with the Osserman problem. It follows immediately from the results in [106] that any self-dual Einstein Walker manifold which is not Ricci flat is $\star$-Einstein, and thus symplectic. Moreover, the almost para-Hermitian Walker structure is never integrable unless it corresponds to a para-complex space form (since the Bochner curvature tensor vanishes due to $W^- = 0$) [61].

Example 5.21.   Four dimensional para-Kaehler manifolds of non-zero constant para-holomorphic sectional curvature are locally described by the para-Kaehler Walker structure J on $\mathcal{M}_{a,b,c}$ (see Theorem 3.14) with

$$a = \alpha x_1^2, \qquad b = \alpha x_2^2, \qquad c = \alpha x_1 x_2\,.$$

The self-dual Weyl curvature operator of this manifold is diagonalizable.

Remark 5.22.   The metrics of Theorem 5.18 (1) can be described in terms of modified Riemannian extensions – see Equation (3.3). Consequently, such a manifold is locally isometric to a metric $g_{D,\phi,T,\mathrm{Id}}$ on the cotangent bundle $T^*\Sigma$ of an affine surface Σ. Here D and ϕ are a torsion free connection and a symmetric $(0,2)$ tensor field chosen arbitrarily on Σ, and T is a multiple of the identity given by $T = \mathcal{B}\,\mathrm{Id}$.

CHAPTER 6

The Spectral Geometry of the Curvature Tensor

6.1 INTRODUCTION

This chapter deals almost exclusively with the signature $(2, 2)$ setting. We introduce the history of the subject in Section 6.2. In Section 6.3, we discuss some general results concerning Jordan Osserman manifolds. Let $\mathcal{M}$ be a Jordan Osserman manifold of signature $(2, 2)$. The possible curvature tensors are classified by Theorem 1.18. If the Jacobi operator is diagonalizable (Type Ia), then $\mathcal{M}$ is locally isometric to a real space form, to a complex space form, or to a para-complex space form [32]. Furthermore, the Jacobi operator can not have complex eigenvalues; thus $\mathcal{M}$ is not Type Ib. If $\mathcal{M}$ is locally symmetric, then either the Jacobi operator is Type Ia or the Jacobi operator is a nilpotent operator of Type II. There are many examples of Jordan Osserman manifolds which are not symmetric and which have nilpotent Jacobi operators of order 2 or 3. Type II Osserman metrics which are not nilpotent are presented.

In Section 6.4, we examine Walker manifolds which are both Osserman and Ivanov–Petrova. We show that such a manifold has either constant sectional curvature or a 2-step nilpotent Jacobi operator. Let $g_{D,\phi}$ be the twisted Riemannian extension given by Equation (3.4). In Section 6.5, we consider the case when M is an affine surface. Affine surfaces with skew symmetric Ricci tensor or with symmetric and degenerate Ricci tensor are studied and Riemannian extensions with commuting curvature operators are analyzed. Finally, curvature commuting Walker 4 dimensional manifolds which are not Riemannian extensions are treated.

6.2 HISTORY

To a large extent, the geometry of a pseudo-Riemannian manifold $\mathcal{M}$ is the study of the curvature $R \in \otimes^4 T^*M$ which is defined by the Levi-Civita connection ∇. Since the whole curvature tensor is difficult to handle, the investigation usually focuses on different objects whose properties allow one to obtain information concerning the full curvature tensor. Any two-point homogeneous space is Osserman and the converse is true in the Riemannian (dim $M \neq 16$) [83, 208, 209] and Lorentzian [30, 128] settings. However, there exist many Osserman pseudo-Riemannian metrics in other signatures [130, 137] which are not symmetric spaces. In particular, the 4 dimensional globally Osserman manifolds are classified except those Osserman metrics in signature $(2, 2)$ whose Jacobi operator is a nilpotent operator [32, 101, 106, 134].

There are many Osserman algebraic curvature tensors which cannot be realized geometrically by an Osserman manifold [32, 136], although they can be realized geometrically at a given point

[130, 137] by a pseudo-Riemannian manifold which is not Osserman. Although the Jacobi operator is probably the most natural operator associated to the curvature tensor, there is some important geometrical information encoded by other operators such as the skew symmetric curvature operator or the higher order Jacobi operator [137]. Moreover, not only the Riemann curvature tensor has been used as a starting object to define curvature operators [129]. The Szabó operator is defined by the covariant derivative of the curvature. The conformal Jacobi operator [33, 37] is defined by the Weyl conformal curvature tensor. Note that any Lorentzian or any odd dimensional Riemannian conformally Osserman manifold is locally conformally flat.

The skew symmetric curvature operator can be regarded as the part of the curvature tensor describing the behavior of circles [171]. Geodesics and circles are classical objects in geometry and physics [3, 4, 18, 186], the latter being preserved by Möbius transformations and thus related to the conformal structure (note that Möbius transformations constitute a special class of conformal transformations characterized by preserving the eigenspaces of the Ricci operator).

6.3   FOUR-DIMENSIONAL OSSERMAN METRICS

The Osserman algebraic curvature tensors are classified in Theorems 1.18 and 1.20. At the geometric level, Type Ia Osserman metrics correspond to real, complex and para-complex space forms, Type Ib Osserman metrics do not exist [32] and Types II and III Osserman metrics with Jacobi operator which is not a nilpotent operator have recently been classified in [106] and [101], respectively. Further, note that any Type II Osserman manifold whose Jacobi operator has non-zero eigenvalues is necessarily a Walker manifold.

6.3.1   OSSERMAN METRICS WITH DIAGONALIZABLE JACOBI OPERATOR

Let $\mathcal{M}$ be a pseudo-Riemannian manifold of signature $(2, 2)$. Then $\mathcal{M}$ is pointwise Osserman if and only if it is Einstein and self-dual or anti-self-dual [8, 157]. Furthermore, Theorem 1.18 shows that any 4 dimensional Jordan Osserman metric is curvature homogeneous.

To decide which Osserman algebraic curvature tensors in Theorem 1.18 are realizable by Osserman manifolds, we consider the equivalent class of Einstein self-dual metrics in dimension 4. In such a case, the curvature operator decomposes as

$$R \equiv \frac{\tau}{12} \, \mathrm{Id}_{\Lambda^2} + \begin{pmatrix} W^+ & 0 \\ 0 & 0 \end{pmatrix},$$

and hence, all information is encoded by the self-dual Weyl curvature operator W^+. Since W^+ is traceless, the non-trivial cases correspond to self-dual Einstein metrics whose self-dual Weyl operator has exactly two or three distinct eigenvalues. (Otherwise W^+ vanishes identically and the metric is Einstein and conformally flat and hence of constant sectional curvature).

Assume the self-dual Weyl curvature operator has exactly two distinct eigenvalues and let Ω denote an eigenvector corresponding to the distinguished eigenvalue. Due to the fact that the induced metric on Λ^+ is of Lorentzian signature, one scales Ω so that $\langle \Omega, \Omega \rangle = \pm 2$. Next, observe that spacelike Ω gives rise to opposite indefinite almost Hermitian structures (g, J) while timelike Ω defines almost para-Hermitian structures $(g, \tilde{J})$. Generalizations of the Goldberg-Sachs Theorem [12, 173] (see also [97]) now show that such structures are indeed locally conformally Kaehler (resp. para-Kaehler) with conformal factor $(24\langle W^+, W^+\rangle)^{1/3}$. The Osserman condition shows this conformal factor is constant. Therefore, the structures associated to Ω are Kaehler (with respect to the opposite orientation) or para-Kaehler, depending on the causality of Ω. Once again, $W^- = 0$ is equivalent to the vanishing of the Bochner curvature tensor [61], and thus the metric corresponds either to an indefinite complex space form or to a para-complex space form.

We use results of [97] to see that the final case where the self-dual Weyl curvature operator has three distinct eigenvalues can not occur. We argue as follows. Let Ω_i be orthogonal eigenvectors of W^+ corresponding to the different eigenvalues λ_i. Since the subbundles $\Lambda^\pm$ are parallel, we may use the second Bianchi identity to see $\delta W^+ = 0$. It now follows that either at least two of the eigenvalues λ_i are equal (which is a contradiction) or otherwise Ω_i's are parallel for all i and thus $(\Omega_1, \Omega_2, \Omega_3)$ defines a hyper-Kaehler or hyper-symplectic structure, which is also a contradiction.

We note that pointwise Osserman manifolds of signature $(2, 2)$ need not be curvature homogeneous; the Jordan normal form of the Jacobi operator may change from point to point [43, 130]. However, if the Jacobi operator of a 4 dimensional Osserman metric has a complex eigenvalue, then the Jordan normal form must remain constant on the whole manifold, and thus $\mathcal{M}$ is curvature homogeneous. Moreover, note that the same occurs for the self-dual Weyl curvature operator.

Curvature homogeneous Einstein metrics of dimension 4 whose Weyl curvature operator is complex diagonalizable have been completely classified [99]. These manifolds are locally homogeneous – they are either locally symmetric, or they are locally isometric to a Lie group with a left invariant metric of a specific type. A remarkable fact of such Lie groups is that the restriction of the curvature operator $R : \Lambda^2 \to \Lambda^2$ to the spaces of self-dual and anti-self-dual forms $R^\pm : \Lambda^\pm \to \Lambda^\pm$ has constant eigenvalues λ, $\lambda e^{2\pi i/3}$, $\lambda e^{4\pi i/3}$ with $\lambda \in \mathbb{R} - \{0\}$. This shows that none of such metrics may be self-dual, and hence no 4 dimensional Osserman metric may have Jacobi operator corresponding to Type Ib.

6.3.2   OSSERMAN WALKER TYPE II METRICS

Assume $\mathcal{M}$ is Jordan Osserman Type II in the classification of Theorem 1.18. Then $\mathcal{M}$ is Ricci flat if and only if $\alpha = \beta = 0$. Moreover, if $\mathcal{M}$ has a Jacobi operator which is not a nilpotent operator, then $\alpha = 4\beta \neq 0$ and $\mathcal{M}$ admits a local parallel field of 2 dimensional planes. Thus, $\mathcal{M}$ is a Walker manifold.

We now discuss joint work with J. C. Díaz-Ramos [106]. We adopt the notation of Example 5.2 to define $\mathcal{M} = \mathcal{M}_{a,b,c}$ and use the curvature classification of Theorem 1.18. We begin with:

Theorem 6.1. *Let $\mathcal{M}$ be a 4 dimensional Type II Jordan Osserman manifold where the Jacobi operator is not a nilpotent operator. Then $\mathcal{M}$ is locally isometric to a manifold $\mathcal{M}_{a,b,c}$ as described in* Example 5.2

where

$$a = x_1^2 \tfrac{\tau}{6} + x_1 P + x_2 Q + \tfrac{6}{\tau}\{Q(T-U) + V(P-V) - 2(Q_4 - V_3)\},$$

$$b = x_2^2 \tfrac{\tau}{6} + x_1 S + x_2 T + \tfrac{6}{\tau}\{S(P-V) + U(T-U) - 2(S_3 - U_4)\},$$

$$c = x_1 x_2 \tfrac{\tau}{6} + x_1 U + x_2 V + \tfrac{6}{\tau}\{-QS + UV + T_3 - U_3 + P_4 - V_4\}.$$

Here P, Q, S, T, U, and V are arbitrary functions of the coordinates (x_3, x_4) and the scalar curvature τ is non-zero.

If $\mathcal{M}$ is Type II Jordan Osserman with Jacobi operator not a nilpotent operator, then $\mathcal{M}$ is locally a Walker manifold, and it is Einstein and either self-dual or anti-self-dual. We adopt the notation of Section 5.3. The anti-self-dual case plays no role so we concentrate on the self-dual case. We integrate the Einstein equations and use Theorem 5.7 and Corollary 5.9 to examine self-dual Walker Osserman metrics. The cases where the scalar curvature vanishes and where it does not vanish are quite different.

Theorem 6.2.   *Let $\mathcal{M} = \mathcal{M}_{a,b,c}$ be as in Example 5.2. Set*

$$\mathcal{E} := \tau^2 - 12\tau W_{11}^+ + 48 \left(W_{12}^+\right)^2.$$

The manifold $\mathcal{M}$ is pointwise Osserman self-dual with $\tau \neq 0$ if and only if there exist functions $\{P, Q, S, T, U, V\}$ of (x_3, x_4) so

$$a = x_1^2 \tfrac{\tau}{6} + x_1 P + x_2 Q + \tfrac{6}{\tau}\{Q(T-U) + V(P-V) - 2(Q_4 - V_3)\},$$
$$b = x_2^2 \tfrac{\tau}{6} + x_1 S + x_2 T + \tfrac{6}{\tau}\{S(P-V) + U(T-U) - 2(S_3 - U_4)\},$$
$$c = x_1 x_2 \tfrac{\tau}{6} + x_1 U + x_2 V + \tfrac{6}{\tau}\{-QS + UV + T_3 - U_3 + P_4 - V_4\}.$$

In this setting

1. *The Jacobi operator of $\mathcal{M}$ has eigenvalues $\{0, \tfrac{\tau}{6}, \tfrac{\tau}{24}, \tfrac{\tau}{24}\}$.*

2. *The Jacobi operator of $\mathcal{M}$ is diagonalizable if and only if $\mathcal{E} = 0$.*

3. *If $\mathcal{E}$ never vanishes, then $\mathcal{M}$ is Jordan Osserman Type II.*

Remark 6.3.   In Theorem 6.22, we will present a result which relates Type II Osserman manifolds with non-zero scalar curvature and modified Riemannian extensions. A geometric interpretation of the differential equations in Theorem 6.4 will be given in Section 6.5.1 where it is shown that such metrics correspond to twisted Riemannian extensions of torsion free connections with skew symmetric Ricci tensor (cf. Theorem 6.20).

Theorem 6.4. *The manifold* $\mathcal{M} = \mathcal{M}_{a,b,c}$ *of Example 5.2 is pointwise Osserman self-dual with* $\tau = 0$ *if and only if there are functions* $\{P, Q, S, T, U, V, \xi, \eta, \gamma\}$ *of* (x_3, x_4) *so*

$$a = x_1 P + x_2 Q + \xi, \quad b = x_1 S + x_2 T + \eta, \quad c = x_1 U + x_2 V + \gamma,$$

where we have the relations:

$$2(Q_4 - V_3) = Q(T - U) + V(P - V),$$
$$2(S_3 - U_4) = S(P - V) + U(T - U),$$
$$T_3 - U_3 + P_4 - V_4 = QS - UV.$$

In this setting, we have:

1. $\mathcal{M}$ *is Type Ia if and only if* $T_3 + U_3 - P_4 - V_4 = 0$ *and* $W_{11}^+ = 0$.

2. $\mathcal{M}$ *is Type II if and only if* $T_3 + U_3 - P_4 - V_4 = 0$ *and* $W_{11}^+ \neq 0$.

3. $\mathcal{M}$ *is Type III if and only if* $T_3 + U_3 - P_4 - V_4 \neq 0$.

We have the following characterization of Jordan Osserman Walker manifolds. Recall that the Jacobi operator of any Jordan Osserman 4 dimensional manifold has either one single eigenvalue $\alpha = \frac{\tau}{12}$ or two distinct eigenvalues α, β such that $\alpha = 4\beta$ [32]. Now, since a necessary condition for a Walker manifold to be anti-self-dual is $\tau = 0$, one has

Theorem 6.5. *Let* $\mathcal{M}$ *be a Jordan Osserman Walker manifold of signature* $(2, 2)$.

1. *If the Jacobi operator is diagonalizable, then either* $\mathcal{M}$ *is flat or* $\mathcal{M}$ *is locally isometric to a para-complex space form.*

2. *If the Jacobi operator is not diagonalizable, then either the Jacobi operator is a 2-step or a 3-step nilpotent operator, or the metric* $g_{a,b,c}$ *is given as in* Theorem 6.1.

Remark 6.6. The Osserman manifolds of Theorem 6.2 are naturally equipped with an almost para-Kaehler structure (cf. Section 5.4 and Theorem 5.18). However, the underlying almost para-complex structure is not integrable unless the Jacobi operator is diagonalizable and the metric is locally isometric to a para-complex space form.

Remark 6.7. Any anti-self-dual Jordan Osserman Walker manifold necessarily has nilpotent Jacobi operator. Despite the fact that many nilpotent Jordan Osserman metrics are known, none of the previous examples are anti-self-dual but all of them correspond to special cases of Theorem 6.4. The general expression of W_{11}^+ in Theorem 5.8 makes it quite intractable and hence it is very difficult to obtain the general form of anti-self-dual Walker manifolds. However, if $a = b = c$ in Example 5.2, anti-self-dual Einstein metrics are characterized by

$$a_{11} = a_{22} = -a_{12}, \qquad a_{13} = a_{14}, \qquad a_{23} = a_{24}, \qquad a_{33} + a_{44} = 2a_{34}.$$

Now, it follows after some calculations that such a metric is given by

$$\begin{aligned} a &= (x_2 - x_1)P(x_2 - x_1, x_3 + x_4) + (x_1 + x_2)\alpha(x_3 + x_4) \\ &+ x_3\beta(x_3 + x_4) + x_4\gamma(x_3 + x_4) + \delta(x_3 + x_4). \end{aligned}$$

This defines an Osserman anti-self-dual Walker manifold whose Jacobi operator vanishes or is a 2-step nilpotent operator, depending on whether $2P_1 + (x_2 - x_1)P_{11}$ is zero or not, for any function P depending on two variables and any single variable functions α, β, γ and δ.

Remark 6.8. Note that any Type III Jordan Osserman Walker manifold is Ricci flat, and thus the Jacobi operator is a 3-step nilpotent operator.

Remark 6.9. The behavior of the eigenvalues is usually the primary object of study when examining Osserman manifolds. However, when the metric under consideration is of indefinite signature, the corresponding eigenspaces are important as well. Indeed, 4 dimensional complex and para-complex space forms have diagonalizable Jacobi operator with eigenvalues $\{\alpha, \frac{\alpha}{4}, \frac{\alpha}{4}\}$ but the eigenspace corresponding to the multiple eigenvalue $\frac{\alpha}{4}$ inherits a definite (positive or negative) metric in the complex case. By contrast, in the para-complex case the induced metric has Lorentzian signature. This is necessarily the case for any Type II Jordan Osserman metric.

Example 6.10. We use the previous analysis to construct Type II Jordan Osserman Walker manifolds which are not nilpotent; we refer to [107] for further details. Let k be a non-zero constant. We specialize the manifolds $\mathcal{M}_{a,b,c}$ of Example 5.2 to define the manifold $\mathcal{M}_{k,f}$ by setting:

$$a = 4kx_1^2 - \tfrac{1}{4k}f(x_4)^2, \quad b = 4kx_2^2, \quad c = 4kx_1x_2 + x_2f(x_4) - \tfrac{1}{4k}f'(x_4).$$

Set

$$\mathcal{E} := 24kf(x_4)f'(x_4)x_2 - 12kf''(x_4)x_1 + 3f(x_4)f''(x_4) + 4f'(x_4)^2.$$

Theorem 6.11. *Adopt the notation of Example 6.10.*

1. *$\mathcal{M}_{k,f}$ is Osserman with eigenvalues $\{0, 4k, k, k\}$.*

2. *If $\mathcal{E}$ vanishes identically, then $\mathcal{M}_{k,f}$ is Type Ia.*

3. If $\mathcal{E}$ never vanishes, then $\mathcal{M}_{k,f}$ is Type II Jordan Osserman.

We note that if $\mathcal{E}$ vanishes at certain points but does not vanish identically, then $\mathcal{M}_{k,f}$ is pointwise Jordan Osserman but not globally Jordan Osserman. In particular, if we take $f(x_4) = x_4$, then $\mathcal{M}_{k,f}$ is Jordan Osserman on the open set $6kx_2x_4 \neq -1$. It is, however, not locally homogeneous since ∇R vanishes only when $x_1 = x_2 = x_4 = 0$. By Theorem 1.13, any Osserman metric is null Osserman. This yields:

Theorem 6.12. *The manifold $\mathcal{M}_{k,f}$ of Example 6.10 is null Osserman with a 2-step nilpotent null Jacobi operator.*

A pseudo-Riemannian manifold $\mathcal{M}$ is said to be *Szabó* if the Szabó operator

$$\mathcal{S}(X) : Y \to \nabla_X \mathcal{R}(Y, X)X$$

has constant eigenvalues on the unit pseudo-sphere bundles $S^{\pm}(\mathcal{M})$ [143] (cf. Section 2.4.2). Any Szabó manifold is locally symmetric in the Riemannian [248] and the Lorentzian [156] setting. In the higher signature setting, there are examples where the Szabó operator is a nilpotent operator and non-trivial [143].

Theorem 6.13. *The manifold $\mathcal{M}_{k,f}$ of Example 6.10 is Szabó but is not Jordan Szabó. If $f' \neq 0$, then $\mathcal{M}_{k,f}$ is not Ivanov–Petrova. The manifold $\mathcal{M}_{k,f}$ is locally symmetric if and only if f' vanishes identically.*

6.4 OSSERMAN AND IVANOV–PETROVA METRICS

The work in this section is also due to E. Calviño-Louzao [75]. An algebraic classification of all curvature models of signature $(2, 2)$ which are Osserman Ivanov–Petrova was given in Section 1.4. This algebraic result leads to the following result in the differentiable setting.

Theorem 6.14. *Let $\mathcal{M}$ have signature $(2, 2)$. Then $\mathcal{M}$ is both pointwise Osserman and pointwise Ivanov–Petrova if and only if either $\mathcal{M}$ is a space of constant sectional curvature or the Jacobi operator of $\mathcal{M}$ either vanishes or is a 2-step nilpotent operator. Moreover, the manifold $\mathcal{M}$ is both Jordan Osserman and Jordan Ivanov–Petrova implies that $\mathcal{M}$ has constant sectional curvature.*

Remark 6.15.   It follows from Theorem 6.14 that rank 4 skew symmetric curvature operators given by (2)–(3) in Theorem 1.24 cannot occur in the differentiable setting and therefore they only exist at the algebraic level.

The class of 4 dimensional pseudo-Riemannian Ivanov–Petrova manifolds can be quite complicated. In [172] the existence of 4 dimensional Ivanov–Petrova manifolds which are not Osserman is shown by means of certain locally conformally flat warped product structures of Robertson-Walker type which are not Einstein. We will construct a large family of Ivanov–Petrova 4 dimensional manifolds with metric of neutral signature which are neither Einstein nor locally conformally flat; many examples which are neither Einstein, nor self-dual nor anti-self-dual will be obtained. We set $c = 0$ to simplify the discussion. We then have:

Theorem 6.16.   *Let $\mathcal{M}_{a,b,c}$ be defined as in Example 5.2 with $c = 0$. If $\mathcal{M}_{a,b,c}$ is Ivanov–Petrova, then $\mathcal{M}_{a,b,c}$ is nilpotent Ivanov–Petrova. Furthermore,*

$$a = x_1 S(x_3, x_4) + A(x_2, x_3, x_4) \text{ and}$$
$$b = x_2 V(x_3, x_4) + B(x_1, x_3, x_4)$$

where $A_{22} B_{11} = 0$.

Remark 6.17.   We apply Theorem 6.16 and consider the special case where

$$a = x_2^2 P(x_3, x_4) + x_1 S(x_3, x_4) + x_2 T(x_3, x_4) + \xi(x_3, x_4),$$
$$b = x_1^2 Q(x_3, x_4) + x_1 U(x_3, x_4) + x_2 V(x_3, x_4) + \eta(x_3, x_4).$$

(6.1)

We shall assume $P \neq 0$ and $Q = 0$; a similar study can be developed assuming $P = 0$ and $Q \neq 0$. By Theorem 5.7, these metrics are *never self-dual*. Furthermore, $\mathcal{M}$ is Ivanov–Petrova if and only if

$$a = x_2^2 P(x_3, x_4) + x_1 S(x_3) + x_2 T(x_3, x_4) + \xi(x_3, x_4),$$
$$b = x_2 V(x_4) + \eta(x_3, x_4),$$

(6.2)

where $P \neq 0$ is a smooth function and S, T, V, ξ and η are arbitrary smooth functions. Moreover, $\mathcal{M}$ defined by Equation (6.2) is Einstein if and only if

$$\eta(x_3, x_4) = \frac{2T_4 - TV}{2P} \quad \text{and} \quad P_4 = PV .$$

Metrics defined by Equation (6.2) provide a large family of 4 dimensional *Ivanov–Petrova manifolds which are neither Einstein nor locally conformally flat*; indeed, they are never self-dual. The analysis of the anti-self-dual condition is really much harder (cf. Theorem 5.8). However, it is possible to specialize those metrics to get examples of 4 dimensional Ivanov–Petrova manifolds which are neither Einstein, nor self-dual or anti-self-dual. In particular, a straightforward calculation using the characterization of the self-dual Weyl curvature operator given in Theorem 5.8 shows that taking

$$a = x_2^2 P(x_3) + x_1 S(x_3) + x_2 T(x_3, x_4) + \xi(x_3, x_4), \quad b = x_2 \kappa + \eta(x_3, x_4),$$

where $P \neq 0$ and $\kappa \neq 0$, we get 4 dimensional *Ivanov–Petrova manifolds which are neither Einstein nor self-dual nor anti-self-dual*.

6.5   RIEMANNIAN EXTENSIONS OF AFFINE SURFACES

The simplest case in Equation (3.4) occurs when M is a surface. Let $M = \Sigma$ be a surface which is equipped with a torsion free connection D. The twisted Riemannian extension $g_{D,\phi}$ defined in Equation (3.4) is then a Walker manifold where, after renumbering the indices appropriately and adopting the notation of Example 5.2, we have

$$\begin{aligned} a &= -2x_1 \Gamma_{33}{}^3(x_3, x_4) - 2x_2 \Gamma_{33}{}^4(x_3, x_4) + \phi_{33}(x_3, x_4), \\ b &= -2x_1 \Gamma_{44}{}^3(x_3, x_4) - 2x_2 \Gamma_{44}{}^4(x_3, x_4) + \phi_{44}(x_3, x_4), \\ c &= -2x_1 \Gamma_{34}{}^3(x_3, x_4) - 2x_2 \Gamma_{34}{}^4(x_3, x_4) + \phi_{34}(x_3, x_4). \end{aligned}$$

Here (x_3, x_4) are local coordinates on Σ. Thus, a Walker manifold is a twisted Riemannian extension in canonical coordinates if and only if a, b, and c are affine functions of (x_1, x_2) where the coefficients are smooth in (x_3, x_4). Theorem 5.5 (4) and Theorem 5.7 show that 4 dimensional twisted Riemannian extensions are self-dual and have a nilpotent Ricci operator.

6.5.1   AFFINE SURFACES WITH SKEW SYMMETRIC RICCI TENSOR
We have the following characterization [129]:

Lemma 6.18.   *Let D be a torsion free connection on a surface Σ. The connection D has a nilpotent Jacobi operator if and only if the Ricci tensor defined by D is skew symmetric.*

Proof. We expand

$$p_\lambda(\mathcal{J}_D(x)) := \det(\mathcal{J}_D(x) - \lambda \operatorname{Id}) = \det(\mathcal{J}_D(x)) - \operatorname{Tr}\{\mathcal{J}_D(x)\}\lambda + \lambda^2 .$$

Since $\mathcal{J}_D(x)x = \mathcal{R}(x, x)x = 0$, we may conclude that

$$\det(\mathcal{J}_D(x)) = 0 .$$

Thus, $\operatorname{Spec}\{\mathcal{J}_D(x)\} = \{0\}$ if and only if

$$\rho_D(x, x) = \operatorname{Tr}\{\mathcal{J}_D(x)\} = 0 .$$

We polarize this identity to see that this condition is equivalent to the condition $\rho_D \in \Lambda^2(T^*\Sigma)$, i.e., D is Ricci anti-symmetric. $\square$

Affine surfaces whose torsion free connection has skew symmetric Ricci tensor have been completely characterized by Derdzinski [100]. He simplified a previous result of Wong [260] to show the following result:

Theorem 6.19.   *For a torsion free connection D on a surface Σ, the Ricci tensor of D is skew symmetric if and only if every point of Σ has a neighborhood U with coordinates (x_3, x_4) in which the only non-vanishing Christoffel symbols of D are*

$$\Gamma_{33}{}^3 = -\partial_{x_3}\varphi \quad and \quad \Gamma_{44}{}^4 = \partial_{x_4}\varphi \quad for \quad \varphi \in C^\infty(\Sigma) \ .$$

Ricci flat self-dual Walker manifolds are described in [100]; this gives a complete description of the metrics in Theorem 6.4:

Theorem 6.20.   *A 4 dimensional Ricci flat self-dual Walker manifold is locally isometric to the cotangent bundle $T^*\Sigma$ of an affine surface (Σ, D) equipped with the twisted Riemannian extension $g_{D,\phi} = g_D + \pi^*\phi$, where D is a torsion free connection with skew symmetric Ricci tensor with Christoffel symbols which may be expressed in adapted coordinates (x_3, x_4) as:*

$$\Gamma_{33}{}^3 = -\partial_{x_3}\varphi \quad and \quad \Gamma_{44}{}^4 = \partial_{x_4}\varphi \ .$$

We adopt the notation of Section 3.5.1 and give an alternative description of self-dual Walker manifolds (cf. Theorem 5.7) in terms of modified Riemannian extensions [77]:

Theorem 6.21.   *Let $\mathcal{M}$ be defined by* Example 5.2. *Then $\mathcal{M}$ is self-dual if and only if it is locally isometric to the cotangent bundle $T^*\Sigma$ of an affine surface (Σ, D), with metric tensor*

$$g = \iota X (\iota \, \mathrm{Id} \circ \iota \, \mathrm{Id}) + \iota \, \mathrm{Id} \circ \iota T + g_D + \pi^*\phi \ .$$

This leads to the following result [77]:

Theorem 6.22.   *Let $\mathcal{M}$ be a Type II Osserman manifold of signature $(2, 2)$ with non-zero scalar curvature τ. There exists an affine surface (Σ, D) so that $\mathcal{M}$ is locally isometric to the cotangent bundle $T^*\Sigma$ with metric tensor*

$$g = \frac{\tau}{6} \cdot \iota \, \mathrm{Id} \circ \iota \, \mathrm{Id} + g_D + \frac{24}{\tau}\pi^*\phi \tag{6.3}$$

where ϕ is the symmetric part of the Ricci tensor of D.

6.5.2 AFFINE SURFACES WITH SYMMETRIC AND DEGENERATE RICCI TENSOR

Any twisted Riemannian extension is necessarily a self-dual Walker manifold. We now investigate some particular cases where the converse also holds [76]. First we specialize Theorem 6.21:

Theorem 6.23. *Let $\mathcal{M}_{a,b,c}$ be defined by* Example 5.2. *Assume $\mathcal{M}_{a,b,c}$ is self-dual. If either $\mathcal{M}_{a,b,c}$ is Ivanov–Petrova or $\mathcal{M}_{a,b,c}$ is Ricci flat, then $\mathcal{M}_{a,b,c}$ is a twisted Riemannian extension given by Equation (3.4).*

Remark 6.24. Ricci flat self-dual Walker manifolds have been investigated in [129]. It was shown that they correspond to twisted Riemannian extensions of torsion free connections with skew symmetric Ricci tensor (see also [100, 184]).

On the other hand, Walker manifolds with a nilpotent Ricci operator are not necessarily twisted Riemannian extensions. For instance, a Walker manifold $\mathcal{M}_{a,b,c}$ with

$$a = 0, \qquad b = x_1 x_2 P(x_3, x_4), \qquad c = \tfrac{1}{2} x_1^2 P(x_3, x_4),$$

has a 2-step nilpotent Ricci operator and is not a twisted Riemannian extension.

Theorem 6.25. *Let (Σ, D) be an affine surface. Then the skew symmetric D-curvature operator is a nilpotent operator if and only if the Ricci tensor is symmetric and degenerate.*

Recall that a tensor field K is said to be *recurrent* if there exists a 1 form σ so $D_X K = \sigma(X) K$ for each vector field X. Further an affine surface (Σ, D) is said to be recurrent if its Ricci tensor is recurrent.

Two affine connections $\bar{D}$ and $\tilde{D}$ are said to be *projectively equivalent* if there is a 1 form σ such that

$$\bar{D}_X Y - \tilde{D}_X Y = \sigma(X) Y + \sigma(Y) X$$

for all vector fields X, Y. Hence, an affine surface (Σ, D) with symmetric Ricci tensor is said to be *projectively flat* if it is locally projectively equivalent to a flat connection.

Theorem 6.26. *Let (Σ, D) be an affine surface with a nilpotent skew symmetric D-curvature operator. Then (Σ, D) is recurrent if and only if around each point there exists a coordinate system (x_3, x_4) in which the non-zero component of D is given by*

$$D_{\partial_{x_3}} \partial_{x_3} = a(x_3, x_4) \partial_{x_4} .$$

Moreover, (Σ, D) is locally symmetric if and only if $a(x_3, x_4) = \alpha\, x_4 + \xi(x_3)$, with $\alpha \in \mathbb{R}$ and ξ a smooth function depending on x_3; (Σ, D) is flat if and only if $\partial_{x_4} a(x_3, x_4) = 0$.

Remark 6.27.   Observe that all locally symmetric connections in Theorem 6.26 are projectively flat.

6.5.3   RIEMANNIAN EXTENSIONS WITH COMMUTING CURVATURE OPERATORS

We refer to [50] for further details concerning the materials we shall present here. Recall the definitions of Section 1.4.5. Let ρ_a^D and ρ_s^D denote the alternating and the symmetric components of the Ricci tensor $\rho^D(x, y) := \text{Tr}\{z \to \mathcal{R}^D(z, x)y\}$:

$$\rho_a^D(x, y) := \tfrac{1}{2}\{\rho^D(x, y) - \rho^D(y, x)\}, \quad \text{and}$$
$$\rho_s^D(x, y) := \tfrac{1}{2}\{\rho^D(x, y) + \rho^D(y, x)\}.$$

Theorem 6.28.   *Let $\mathcal{M}$ be the twisted Riemannian extension $g_{D,\phi}$ given by Equation (3.4) of an affine surface (Σ, D); we impose no additional restrictions on D. Then:*

1. *$\rho_a^D = 0$ if and only if $\mathcal{M}$ is curvature–curvature commuting.*

2. *$\rho_s^D = 0$ if and only if $\mathcal{M}$ is Osserman.*

3. *Either $\rho_a^D = 0$ or $\rho_s^D = 0$ if and only if $\mathcal{M}$ is Jacobi–Ricci commuting.*

4. *$\rho_D = 0$ if and only if $\mathcal{M}$ is Jacobi–Jacobi commuting.*

Remark 6.29.   If D is the torsion free connection on $\mathbb{R}^2$ with non-zero Christoffel symbols

$$D_{\partial_{x_3}} \partial_{x_4} = D_{\partial_{x_4}} \partial_{x_3} = f(x_3)\partial_{x_3}, \quad D_{\partial_{x_4}} \partial_{x_4} = f(x_3)\partial_{x_4} \ ,$$

for $f = f(x_3)$ with $f'(x_3) \neq 0$, we have $\rho_D^s = 0$ while $\rho_D^a \neq 0$. Similarly, if we have that

$$D_{\partial_{x_3}} \partial_{x_3} = f(x_3, x_4)\partial_{x_4} \ ,$$

for $f = f(x_3, x_4)$ with $f_4 \neq 0$, it follows that $\rho_D^s \neq 0$ while $\rho_D^a = 0$. Thus, the four possibilities in Theorem 6.28 are distinct.

6.5.4 OTHER EXAMPLES WITH COMMUTING CURVATURE OPERATORS

We adopt the notation of Example 5.2. Consider the Walker manifold $\mathcal{M}_{a,b,c}$ defined by

$$a = \alpha x_1 x_2, \quad b = -\alpha x_1 x_2, \quad c = \alpha(x_2^2 - x_1^2)/2.$$

This manifold is a locally symmetric Walker manifold which is Jacobi–Ricci commuting and for which $\rho^2 = -\alpha^2$ Id. We refer to [150] for further details. This manifold satisfies the condition

$$R(\rho x, y, z, w) = R(x, y, \rho z, w) .$$

If we let ρ give $\mathcal{M}$ a pseudo-Hermitian structure, then the curvature tensor of $\mathcal{M}$ lies in the decomposition factor $\mathcal{W}_7$ of the Tricerri-Vanhecke curvature decomposition of Theorem 1.8. In particular, it does not satisfy the Gray identity and this almost complex structure is not integrable.

A complete classification of manifolds satisfying curvature commuting conditions is available in certain special cases. We refer to [150] for the proof of the following result:

Theorem 6.30. *Let $\mathcal{M}_{a,b,c}$ be the Walker manifold defined by Example 5.2 with $a = b = 0$. Then $\mathcal{M}_{a,b,c}$ is Jacobi–Ricci commuting if and only if*

$$c = x_1 P(x_3, x_4) + x_2 Q(x_3, x_4) + S(x_3, x_4)$$

where one of the following two conditions holds:

1. *$P_3 = Q_4$, i.e., $Q dx_3 + P dx_4$ is a closed 1 form.*

2. *There exist $(u, v, w) \neq (0, 0, 0)$ so*

$$P = \frac{w}{u + v x_3 + w x_4} \quad and \quad Q = \frac{v}{u + v x_3 + w x_4} .$$

CHAPTER 7

Hermitian Geometry

7.1 INTRODUCTION

In Hermitian geometry, one often examines the relationship between the properties of the structure (g, J) and properties of the curvature tensor. The Goldberg Conjecture [160] is an example; it is conjectured that compact almost Kaehler Riemannian Einstein manifolds are necessarily Kaehler. The Goldberg conjecture is global in nature and is still open. However, additional curvature conditions are known which locally imply the integrability of the almost complex structure. Proofs in this area make use of relations involving the curvature terms. For example, in the almost Kaehler context, the relation

$$\tau - \tau^\star = \tfrac{1}{2}|\nabla\Omega|^2$$

implies that $|\nabla\Omega|^2 = 0$ if $\tau = \tau^\star$; this means that (g, J) is *isotropic Kaehler*. In the Riemannian setting, this shows $\nabla\Omega = 0$ so (g, J) is in fact Kaehler and thereby establishes the desired integrability condition [233]. In the higher signature setting, the identity $|\nabla\Omega|^2 = 0$ does not imply that $\nabla\Omega = 0$. Thus, the class of isotropic Kaehler structures properly contains the class of Kaehler structures [28, 131]. Further historical information is given in Section 7.2.

In Chapter 7, we will construct examples of Walker manifolds which illustrate that indefinite almost Hermitian structures exhibit very different behaviors than in the Riemannian setting. In particular, we will construct manifolds which admit isotropic Kaehler structures which are not Kaehler. We shall examine proper almost complex structures on Walker manifolds following the discussion in [198] to obtain a local description of those metrics which are Hermitian, almost Kaehler, self-dual, $\star$-Einstein or Einstein.

In Section 7.3, we will show that any proper almost Hermitian structure on a 4 dimensional Walker manifold is isotropic Kaehler. We will study proper hyper-para-Hermitian structures and the $\star$-Einstein equation. Proper Hermitian Walker structures are studied in Section 7.4, where the Einstein equation of Theorem 5.5 (4) is explicitly integrated. This allows the construction of examples of indefinite Einstein strictly almost Hermitian structures showing that the integrability result given by Kirchberg [180] does not hold for metrics of signature $(2, 2)$. Special attention is paid to locally conformally Kaehler structures when we discuss Vaisman manifolds. Almost Kaehler proper structures are treated in Section 7.5, with special attention to the $\star$-Einstein and the Einstein cases. Almost Kaehler proper structures of constant sectional curvature are considered when we construct local examples of strictly almost Kaehler Einstein structures which are not Kaehler.

7.2   HISTORY

A well-known phenomena in geometry is the incompatibility between certain additional structures (Hermitian, quaternionic, contact, etc.) and the curvature of the underlying metric. As we have shown in Chapter 5, the curvature influences the underlying structure of the manifold by equipping it with some extra properties. This is the spirit behind results like the Goldberg-Sachs Theorem [12, 14] and related results [84, 97].

The best examples of the antagonism between various geometric properties are to be found within the context of Kaehler geometry. Many of these results have been obtained by making use of Bochner's type formulae which relate curvature objects with invariants of the structure under consideration [167]. Sometimes the explicit integration of the Chern classes provides the desired information, as occurs for compact Einstein almost Kaehler 4 dimensional manifolds [233, 234]. In a general setting, these integral formulae imply that some scalar invariants of the structure vanish. For instance, one of the candidates for vanishing is $|\nabla J|$ since, in the Riemannian setting, $|\nabla J| = 0$ implies $\nabla J = 0$. However, in the pseudo-Riemannian context, the tensor ∇J may be null but non-zero; this makes the class of isotropic structures (i.e., those where $|\nabla J| = 0$, but $\nabla J \neq 0$) of special interest.

The first examples of isotropic Kaehler structures were found in [131], where compact 4 dimensional examples of signature $(2, 2)$ were exhibited – see also [28] for 6 dimensional examples. Later, it was shown [93, 94] that isotropy is a common property in the Walker setting as any proper almost Hermitian structure is isotropic Kaehler (see Section 7.3).

Finally, it is worth mentioning that the use of integral formulae extends (under suitable conditions) to the indefinite setting, as shown in [238], where a previously used integral expression is specialized to the case of pseudo-Riemannian compact almost Kaehler Einstein geometry and explicitly confirmed for a neutral signature metric on the 8-torus in [199]. This involves a generalization of the isotropy condition to k^{th} order isotropic Kaehler: $\nabla^k J$ is non-zero but null. Many open problems arise now in the indefinite setting: is a compact, almost Kaehler Einstein manifold isotropic Kaehler, or at least k^{th} order isotropic Kaehler for some k? One also has a related question: is there a k such that a pseudo-Riemannian, almost Hermitian manifold which is l^{th} order isotropic Kaehler for $1 \leq l \leq k$ is in fact Kaehler?

7.3   ALMOST HERMITIAN GEOMETRY OF WALKER MANIFOLDS

Recall that an indefinite almost Hermitian structure (g, J) is said to be an *isotropic Kaehler* structure if one has that $|\nabla J|^2 = 0$ (or equivalently that $|\nabla \Omega|^2 = 0$) but $\nabla J \neq 0$. Examples of isotropic Kaehler structures have been given first in [131] in dimension 4 and subsequently in [28] in dimension 6.

7.3.1 THE PROPER ALMOST HERMITIAN STRUCTURE OF A WALKER MANIFOLD

We follow the discussion in [198].

Example 7.1. Let $\mathcal{M}_{a,b,c} := (\mathcal{O}, g_{a,b,c})$ be the pseudo-Riemannian manifold of Example 5.2. We take the induced orthonormal basis of Equation (5.1):

$$e_1 := \tfrac{1}{2}(1-a)\partial_{x_1} + \partial_{x_3}, \qquad e_2 := -c\partial_{x_1} + \tfrac{1}{2}(1-b)\partial_{x_2} + \partial_{x_4},$$

$$e_3 := -\tfrac{1}{2}(1+a)\partial_{x_1} + \partial_{x_3}, \qquad e_4 := -c\partial_{x_1} - \tfrac{1}{2}(1+b)\partial_{x_2} + \partial_{x_4}.$$

We shall let $\mathcal{C}_{a,b,c} := (\mathcal{O}, g_{a,b,c}, J)$ where J is the *proper almost complex structure* defined by:

$$J = e_2 \otimes e^1 - e_1 \otimes e^2 + e_4 \otimes e^3 - e_3 \otimes e^4 \ .$$

This means that

$$J : e_1 = e_2, \quad J e_2 = -e_1, \quad J e_3 = e_4, \quad J e_4 = -e_3 \,.$$

The map J induces a positive $\frac{\pi}{2}$-rotation on the degenerate parallel field $\mathcal{D} := \mathrm{Span}\{\partial_{x_1}, \partial_{x_2}\}$:

$$J\partial_{x_1} = \partial_{x_2}, \qquad J\partial_{x_3} = -c\partial_{x_1} + \tfrac{1}{2}(a-b)\partial_{x_2} + \partial_{x_4},$$

$$J\partial_{x_2} = -\partial_{x_1}, \qquad J\partial_{x_4} = \tfrac{1}{2}(a-b)\partial_{x_1} + c\partial_{x_2} - \partial_{x_3}.$$

The following result shows that the class of isotropic Kaehler structures is larger than might at first sight be expected:

Theorem 7.2. *Let $\mathcal{C}_{a,b,c}$ be as given in* Example 7.1.

1. $\mathcal{C}_{a,b,c}$ *is almost Kaehler if and only if $a_1 + b_1 = 0$ and $a_2 + b_2 = 0$.*

2. $\mathcal{C}_{a,b,c}$ *is Hermitian if and only if $a_1 - b_1 = 2c_2$ and $a_2 - b_2 = -2c_1$.*

3. $\mathcal{C}_{a,b,c}$ *is Kaehler if and only if $a_1 = -b_1 = c_2$ and $a_2 = -b_2 = -c_1$.*

4. *One has that $|\nabla\Omega|^2 = 0$, $|d\Omega|^2 = 0$, $|\delta\Omega|^2 = 0$, and $|N_J|^2 = 0$.*

5. $\tau = a_{11} + b_{22} + 2c_{12}$ *and* $\tau^\star = -a_{22} - b_{11} + 2c_{12}$.

Remark 7.3. Examples of compact isotropic Kaehler structures can be constructed on tori by taking a, b and c to be periodic functions on $\mathbb{R}^4$. Moreover, note that in the general situation the isotropic Kaehler structures given by Example 7.1 are neither complex nor symplectic (cf. Theorem 7.2). Hence, for special choices of functions (which may still be assumed to be periodic) satisfying the conditions in Theorem 7.2, examples of symplectic or integrable isotropic Kaehler structures can be given.

Remark 7.4. The $\star$-Einstein equations $(\rho_0^\star = \rho^\star - \frac{\tau^\star}{4} g = 0)$ of $\mathcal{C}_{a,b,c}$ can be written as a system of partial differential equations as follows:

$$(\rho_0^\star)_{13} = -(\rho_0^\star)_{24} = -(\rho_0^\star)_{31} = (\rho_0^\star)_{42} = \tfrac{1}{4}(a_{22} - b_{11}) = 0,$$

$$(\rho_0^\star)_{14} = -(\rho_0^\star)_{32} = -\tfrac{1}{2}(a_{12} - c_{11}) = 0,$$

$$(\rho_0^\star)_{23} = -(\rho_0^\star)_{41} = -\tfrac{1}{2}(b_{12} - c_{22}) = 0,$$

$$(\rho_0^\star)_{33} = \tfrac{1}{4}\big\{a_1 b_1 + a_2(b_2 - c_1) + b_2 c_1 + c_2(a_1 - b_1) - c_1^2 - c_2^2 + 2c(a_{12} - c_{11}) \\ + b a_{22} - 2a_{24} + a b_{11} - 2b_{13} - (a+(b)\, c_{12} + 2c_{14} + 2c_{23}\big\} = 0,$$

$$(\rho_0^\star)_{34} = -\tfrac{1}{4}\{(a-b)(a_{12} - c_{11}) + c(a_{22} - b_{11})\} = 0,$$

$$(\rho_0^\star)_{43} = \tfrac{1}{4}\{(a-b)(b_{12} - c_{22}) + c(a_{22} - b_{11})\} = 0,$$

$$(\rho_0^\star)_{44} = \tfrac{1}{4}\big\{a_1 b_1 + a_2(b_2 - c_1) + b_2 c_1 + c_2(a_1 - b_1) - c_1^2 - c_2^2 + 2c(b_{12} - c_{22}) \\ + b a_{22} - 2a_{24} + a b_{11} - 2b_{13} - (a+(b)\, c_{12} + 2c_{14} + 2c_{23}\big\} = 0.$$

7.3.2   PROPER ALMOST HYPER-PARA-HERMITIAN STRUCTURES

Example 7.5. The notion of an almost hyper-para-Hermitian structure was defined previously in Section 2.5.2. Let $\mathcal{M}_{a,b,c}$ be as in Example 5.2. We adopt the notation of Equation (5.1) to define an orthonormal frame $\{e_i\}_{i=1}^4$ of the tangent bundle of $\mathbb{R}^4$. We use Equation (1.12) to define a corresponding orthonormal frame $\{E_j^\pm\}_{j=1}^3$ for $\Lambda^2 = \Lambda^+ \oplus \Lambda^-$:

$$E_1^\pm = \tfrac{1}{\sqrt{2}}\{e^1 \wedge e^2 \pm e^3 \wedge e^4\}, \quad E_2^\pm = \tfrac{1}{\sqrt{2}}\{e^1 \wedge e^3 \pm e^2 \wedge e^4\},$$
$$E_3^\pm = \tfrac{1}{\sqrt{2}}\{e^1 \wedge e^4 \mp e^2 \wedge e^3\}.$$

The bivectors $\{E_j^-\}_{j=1}^3$ define endomorphisms $\{J_j\}_{j=1}^3$ of $T\mathbb{R}^4$ such that

$$J_1^2 = -1, \quad J_2^2 = J_3^2 = 1, \quad J_1 J_2 = -J_2 J_1 = J_3.$$

We shall say that the almost hyper-para-complex structure $\mathcal{J} := \{J_1, J_2, J_3\}$ which is defined by means of E_1^-, E_2^-, E_3^- is *proper*. Note that J_1 is an isometry of the Walker manifold $g_{a,b,c}$ while J_2, J_3 are anti-isometries, i.e., $\mathcal{J}$ is an almost hyper-para-Hermitian structure. Let

$$\mathcal{H}_{a,b,c} := (\mathcal{O}, g_{a,b,c}, \mathcal{J}).$$

From Equations (5.1) and (1.12) one gets the description in terms of the coordinate frame:

$$
J_1 = \begin{pmatrix} 0 & -1 & -c & \frac{a-b}{2} \\ 1 & 0 & \frac{a-b}{2} & c \\ 0 & 0 & 0 & -1 \\ 0 & 0 & 1 & 0 \end{pmatrix}, \quad
J_2 = \begin{pmatrix} 1 & 0 & a & 0 \\ 0 & -1 & 0 & -b \\ 0 & 0 & -1 & 0 \\ 0 & 0 & 0 & 1 \end{pmatrix}, \quad
J_3 = \begin{pmatrix} 0 & 1 & c & \frac{a+b}{2} \\ 1 & 0 & \frac{a+b}{2} & c \\ 0 & 0 & 0 & -1 \\ 0 & 0 & -1 & 0 \end{pmatrix}.
$$

Note that J_1 is the proper almost complex structure defined in Example 7.1.

Theorem 7.6. *Let $\mathcal{H}_{a,b,c}$ be as in Example 7.5.*

1. $\mathcal{J} := \{J_1, J_2, J_3\}$ *is integrable if and only if*

$$a = x_1^2 \mathcal{B}(x_3, x_4) + x_1 P(x_3, x_4) + \xi(x_3, x_4),$$

$$b = x_2^2 \mathcal{B}(x_3, x_4) + x_2 T(x_3, x_4) + \eta(x_3, x_4),$$

$$c = x_1 x_2 \mathcal{B}(x_3, x_4) + \tfrac{1}{2} x_1 T(x_3, x_4) + \tfrac{1}{2} x_2 P(x_3, x_4) + \gamma(x_3, x_4).$$

2. $\mathcal{J}$ *is hyper-para-Kaehler if and only if $a = a(x_3, x_4)$, $b = b(x_3, x_4)$, and $c = c(x_3, x_4)$.*

Remark 7.7. Observe that all metrics given in Theorem 7.6 (1) are self-dual (see Theorem 5.7) and moreover, the Ricci operator has a unique eigenvalue $\lambda = \frac{3}{2}\mathcal{B}(x_3, x_4)$, which is a double root of its minimal polynomial. We take $\mathcal{B} = \alpha$ to be an arbitrary constant and we recover Example 5.21 by taking $P = T = \xi = \eta = \gamma = 0$.

Results of [178] show that any hyper-para-Kaehler structure is Ricci flat, just as in the definite case [46]. Neutral signature Ricci flat non-flat metrics on complex tori and primary Kodaira surfaces have been constructed in [223]. These metrics are induced by proper hyper-para-Kaehler Walker structures on $\mathbb{R}^4$. Further observe that proper hyper-para-Kaehler structures correspond to Walker manifolds admitting two orthogonal null parallel vector fields $\{\partial_{x_1}, \partial_{x_2}\}$ – see Theorem 3.2.

7.4   HERMITIAN WALKER MANIFOLDS OF DIMENSION FOUR

This section contains joint work with J. Davidov, J. C. Díaz-Ramos, Y. Matsushita, and O. Muškarov [94]. Throughout this section, let $\mathcal{C}_{a,b,c}$ be defined as in Example 7.1.

7.4.1   PROPER HERMITIAN WALKER STRUCTURES

We shall investigate curvature properties of $\mathcal{M}_{a,b,c}$ by considering the associated proper structure $\mathcal{C}_{a,b,c}$. It turns out that this structure exhibits a very rich behavior providing examples, as we mentioned in the previous section, of indefinite Ricci flat (non-flat) Kaehler structures on tori and primary Kodaira surfaces [223] as well as flat non-Kaehler almost Kaehler structures [93]. This is in sharp contrast to the Riemannian case and it is important to recognize that such an exceptional behavior comes from the fact that any proper almost Hermitian Walker structure is isotropic Kaehler but not necessarily Kaehler (cf. Theorem 7.2).

Theorem 7.8. $\mathcal{C}_{a,b,c}$ *is Hermitian and self-dual if and only if*

$$a = x_1^2 \mathcal{B}(x_3, x_4) + x_1 x_2 \mathcal{D}(x_3, x_4) + x_1 P(x_3, x_4) + x_2 Q(x_3, x_4) + \xi(x_3, x_4),$$

$$b = x_2^2 \mathcal{B}(x_3, x_4) - x_1 x_2 \mathcal{D}(x_3, x_4) + x_1 S(x_3, x_4) + x_2 T(x_3, x_4) + \eta(x_3, x_4),$$

$$c = \tfrac{1}{2}(x_2^2 - x_1^2)\mathcal{D}(x_3, x_4) + x_1 x_2 \mathcal{B}(x_3, x_4) - \tfrac{1}{2}x_1\{Q(x_3, x_4) - T(x_3, x_4)\}$$
$$+ \tfrac{1}{2}x_2\{P(x_3, x_4) - S(x_3, x_4)\} + \gamma(x_3, x_4).$$

Remark 7.9.   Assume that $\mathcal{C}_{a,b,c}$ is Hermitian and self-dual. One can then use Theorem 7.8 to see that the Ricci operator has complex eigenvalues

$$\lambda = \tfrac{3}{2}\mathcal{B}(x_3, x_4) \pm \sqrt{-1}\mathcal{D}(x_3, x_4) \quad \text{of multiplicity two .}$$

The $\star$-Einstein equations were discussed in Remark 7.4.

Theorem 7.10. $\mathcal{C}_{a,b,c}$ *is Hermitian and $\star$-Einstein if and only if the functions a, b, c have one of the following three forms where $\kappa \neq 0$ is constant.*

1. *Form I ($\tau^\star = 8\kappa$):*

$$a = \kappa(x_1^2 - x_2^2) + x_1 P(x_3, x_4) + x_2 Q(x_3, x_4) + \xi(x_3, x_4),$$

$$b = \kappa(x_2^2 - x_1^2) - x_1 P(x_3, x_4) - x_2 Q(x_3, x_4) - \xi(x_3, x_4) + \tfrac{1}{\kappa}\{P_3(x_3, x_4) - Q_4(x_3, x_4)\},$$

$$c = 2\kappa x_1 x_2 - x_1 Q(x_3, x_4) + x_2 P(x_3, x_4) + \gamma(x_3, x_4).$$

2. *Form II ($\tau^\star = 2\kappa$):*

$$a = \kappa x_1^2 + x_1 P(x_3, x_4) + x_2 Q(x_3, x_4) + \xi(x_3, x_4),$$

$$b = \kappa x_2^2 + x_1 S(x_3, x_4) + x_2 T(x_3, x_4) - \xi(x_3, x_4)$$

$$- \tfrac{1}{4\kappa}\{4(3S_3(x_3, x_4) - P_3(x_3, x_4) + 3Q_4(x_3, x_4) - T_4(x_3, x_4))$$

$$-(P(x_3, x_4) + S(x_3, x_4))^2 - (Q(x_3, x_4) + T(x_3, x_4))^2\},$$

$$c = \kappa x_1 x_2 - \tfrac{1}{2} x_1 (Q(x_3, x_4) - T(x_3, x_4)) + \tfrac{1}{2} x_2 (P(x_3, x_4) - S(x_3, x_4)) + \gamma(x_3, x_4).$$

3. *Form III* $(\tau^\star = 0)$:

$$a = x_1 P(x_3, x_4) + x_2 Q(x_3, x_4) + \xi(x_3, x_4),$$

$$b = x_1 S(x_3, x_4) + x_2 T(x_3, x_4) + \eta(x_3, x_4),$$

$$c = -\tfrac{1}{2} x_1 (Q(x_3, x_4) - T(x_3, x_4)) + \tfrac{1}{2} x_2 (P(x_3, x_4) - S(x_3, x_4)) + \gamma(x_3, x_4),$$

where $4(3S_3 - P_3 + 3Q_4 - T_4) = (P + S)^2 + (Q + T)^2.$

We use the results of Section 5.3 to see:

Theorem 7.11. $\mathcal{C}_{a,b,c}$ *is Hermitian and Einstein if and only if the functions a, b, c have one of the following three forms where* $\kappa \neq 0$ *is constant:*

1. *Form I* $(\tau = 8\kappa)$:

$$a = \kappa(x_1^2 - x_2^2) + x_1 P(x_3, x_4) + x_2 Q(x_3, x_4) + \xi(x_3, x_4),$$

$$b = \kappa(x_2^2 - x_1^2) - x_1 P(x_3, x_4) - x_2 Q(x_3, x_4) - \xi(x_3, x_4) + \tfrac{1}{\kappa}\{P_3(x_3, x_4) - Q_4(x_3, x_4)\},$$

$$c = 2\kappa x_1 x_2 - x_1 Q(x_3, x_4) + x_2 P(x_3, x_4) + \gamma(x_3, x_4).$$

2. *Form II* $(\tau = 6\kappa)$:

$$a = \kappa x_1^2 + x_1 P(x_3, x_4) + x_2 Q(x_3, x_4)$$
$$+ \tfrac{1}{4\kappa}\{4(P_3(x_3, x_4) - S_3(x_3, x_4)) - 8Q_4(x_3, x_4) + 2Q(x_3, x_4)(Q(x_3, x_4)$$
$$+ T(x_3, x_4)) + P(x_3, x_4)^2 - S(x_3, x_4)^2\},$$

$$b = \kappa x_2^2 + x_1 S(x_3, x_4) + x_2 T(x_3, x_4) + \tfrac{1}{4\kappa}\{-4(Q_4(x_3, x_4) - T_4(x_3, x_4)) - 8S_3(x_3, x_4)$$
$$+ 2S(x_3, x_4)(P(x_3, x_4) + S(x_3, x_4)) - (Q(x_3, x_4)^2 - T(x_3, x_4)^2)\},$$

$$c = \kappa x_1 x_2 - \tfrac{1}{2} x_1 (Q(x_3, x_4) - T(x_3, x_4)) + \tfrac{1}{2} x_2 (P(x_3, x_4) - S(x_3, x_4))$$
$$+ \tfrac{1}{4\kappa}\{2(P_4(x_3, x_4) + S_4(x_3, x_4)) + 2(Q_3(x_3, x_4) + T_3(x_3, x_4))$$
$$+ T(x_3, x_4)(P(x_3, x_4) - S(x_3, x_4)) - Q(x_3, x_4)(P(x_3, x_4) + 3S(x_3, x_4))\}.$$

3. *Form III* $(\tau = 0)$:

$$a = x_1 P(x_3, x_4) + x_2 Q(x_3, x_4) + \xi(x_3, x_4),$$

$$b = x_1 S(x_3, x_4) + x_2 T(x_3, x_4) + \eta(x_3, x_4),$$

$$c = -\tfrac{1}{2} x_1 (Q(x_3, x_4) - T(x_3, x_4)) + \tfrac{1}{2} x_2 (P(x_3, x_4) - S(x_3, x_4)) + \gamma(x_3, x_4),$$

where

$$8Q_4 - 4(P_3 - S_3) = 2Q(Q + T) + (P^2 - S^2),$$

$$8S_3 + 4(Q_4 - T_4) = 2S(P + S) - (Q^2 - T^2),$$

$$2(P_4 + S_4) + 2(Q_3 + T_3) = Q(P + 3S) - T(P - S).$$

Remark 7.12. Metrics of the form given in Theorem 7.11 (3) have received attention in the literature of Osserman manifolds. Since any 4 dimensional twisted Riemannian extension $g_{D,\phi}$ given by Equation (3.4) is self-dual [76], metrics of this form are Osserman with nilpotent Jacobi operator and moreover they are realized as twisted Riemannian extensions of torsion free connections with skew symmetric Ricci tensor [106, 129].

Corollary 7.13. *If $\mathcal{C}_{a,b,c}$ is Hermitian and Einstein, then $\mathcal{C}_{a,b,c}$ is $\star$-Einstein.*

Remark 7.14. $\mathcal{C}_{a,b,c}$ is Hermitian, self-dual and Einstein if and only if the functions a, b and c are of the form II or III in Theorem 7.11.

Theorem 3.2 of [180] shows that an almost Hermitian manifold (M, g, J) of dimension 4 with positive definite metric is Hermitian if the metric g is Einstein and if the Weyl curvature satisfies

$$|W^+|^2 = \tfrac{1}{24}(\tau - 3\tau^\star)^2 \neq 0$$

at every point of M. We shall give an example showing that the analogous result is not true in the signature $(2, 2)$ setting. If M has signature $(2, 2)$, the role of W^+ is played by W^- since the almost complex structures compatible with the metric and the orientation are parametrized by sections of Λ^-. Note also that the metric on Λ^2 in [180] is one half of the metric used here. Thus, in our situation, the analogous condition for W^+ is

$$|W^-|^2 = \tfrac{1}{96}(\tau - 3\tau^\star)^2 \neq 0 .$$

Example 7.15. Let $\mathcal{C}_{a,b,c}$ be defined by $a = x_1^2 + \tfrac{2}{3}x_2^2 + \tfrac{2}{\sqrt{3}}x_1 x_2$, $b = x_2^2$, and $c = -\tfrac{1}{\sqrt{3}}x_2^2$. A straightforward calculation shows that $\mathcal{C}_{a,b,c}$ is Einstein, that $\tau = 4$, that $\tau^\star = -\tfrac{4}{3}$, and that

$$W^- = \begin{pmatrix} \dfrac{2}{3} & -\dfrac{1}{\sqrt{3}} & -\dfrac{1}{3} \\[2mm] \dfrac{1}{\sqrt{3}} & -\dfrac{2}{3} & -\dfrac{1}{\sqrt{3}} \\[2mm] \dfrac{1}{3} & -\dfrac{1}{\sqrt{3}} & 0 \end{pmatrix} \quad \text{so} \quad |W^-|^2 = \tfrac{2}{3} = \tfrac{1}{96}(\tau - 3\tau^\star)^2 .$$

On the other hand, it follows from Theorem 7.2 that the proper almost complex structure J is not integrable. Note that W^- has degenerate spectrum $\{-\frac{2}{3}, \frac{1}{3}, \frac{1}{3}\}$ and that the eigenspace corresponding to the simple eigenvalue is timelike (and hence it determines an integrable almost para-Hermitian structure [173]).

7.4.2   LOCALLY CONFORMALLY KAEHLER STRUCTURES

Recall that an indefinite Hermitian manifold (M, g, J) is said to be *locally conformally Kaehler* if for any point $P \in M$ there exists an open neighborhood U and a function $f : U \to \mathbb{R}$ such that $(U, e^{2f}g, J)$ is an indefinite Kaehler manifold [109]. Recall that a Hermitian structure (g, J) is locally conformally Kaehler if and only if the Lee form ω defined by $d\Omega = \omega \wedge \Omega$ is a closed 1 form [110]. In particular, the Lee form ω of the manifold $\mathcal{C}_{a,b,c}$ of Example 7.1 is given by:

$$\omega = -\tfrac{1}{2}(a_1 + b_1)\, dx_3 - \tfrac{1}{2}(a_2 + b_2) dx_4 \,. \tag{7.1}$$

Theorem 7.16. *$\mathcal{C}_{a,b,c}$ is locally conformally Kaehler if and only if*

$$a = x_1 P(x_3, x_4) + x_2 Q(x_3, x_4) + \mathcal{A}(x_1, x_2, x_3, x_4),$$

$$b = x_1 P(x_3, x_4) + x_2 Q(x_3, x_4) + \eta(x_3, x_4) - \mathcal{A}(x_1, x_2, x_3, x_4),$$

$$c = \mathcal{B}(x_1, x_2, x_3, x_4).$$

Here the function $\mathcal{A} + \sqrt{-1}\mathcal{B}$ is holomorphic with respect to $w = x_1 + \sqrt{-1}x_2$ and $P_4 = Q_3$. Moreover, $\mathcal{C}_{a,b,c}$ is Kaehler if and only if $P = Q = 0$.

Remark 7.17.   The metrically equivalent vector field B of the Lee form ω is called the *Lee vector field* and JB is usually named as the *anti-Lee vector field*. A special class of locally conformally Kaehler structures, the so-called *Vaisman manifolds*, corresponds to the case of parallel Lee form. Note that although proper locally conformally Kaehler structures are not necessarily Vaisman structures, the distribution generated by B and JB is parallel since

$$B = -P(x_3, x_4)\partial_{x_1} - Q(x_3, x_4)\partial_{x_2} \,.$$

Further observe that many of the striking differences between positive definite locally conformally Kaehler structures and the indefinite counterpart lie on the fact that the distribution $\{B, JB\}$ may be degenerate, which indeed holds in the case under consideration (see [109]).

Theorem 7.18.   *Assume the Lee form ω of $\mathcal{C}_{a,b,c}$ is nowhere vanishing. Then ω is parallel if and only if*

$$a = -2x_1(P(x_3, x_4)P_3(x_3, x_4) + Q(x_3, x_4)Q_3(x_3, x_4))\mathcal{E}(x_3, x_4)$$
$$+2x_2(P(x_3, x_4)Q_3(x_3, x_4) - Q(x_3, x_4)P_3(x_3, x_4))\mathcal{E}(x_3, x_4) + \xi(x_3, x_4),$$
$$b = 2x_1(Q(x_3, x_4)P_4(x_3, x_4) - P(x_3, x_4)Q_4(x_3, x_4))\mathcal{E}(x_3, x_4)$$
$$-2x_2(P(x_3, x_4)P_4(x_3, x_4) + Q(x_3, x_4)Q_4(x_3, x_4))\mathcal{E}(x_3, x_4) + \eta(x_3, x_4),$$
$$c = x_1(Q(x_3, x_4)(P_3(x_3, x_4) - Q_4(x_3, x_4)) - 2P(x_3, x_4)Q_3(x_3, x_4))\mathcal{E}(x_3, x_4)$$
$$-x_2(P(x_3, x_4)(P_3(x_3, x_4) - Q_4(x_3, x_4)) + 2Q(x_3, x_4)Q_3(x_3, x_4))\mathcal{E}(x_3, x_4)$$
$$+\gamma(x_3, x_4),$$

where $P_4 - Q_3 = 0$ and $\mathcal{E} := (P^2 + Q^2)^{-1}$ satisfies $P_3 + Q_4 = -\mathcal{E}^{-1}$.

Example 7.19. Let $\mathcal{E}(x_3, x_4) := (px_3 + qx_4 + r)^{-1}$. We solve the relations of Theorem 7.18 by setting $P(x_3, x_4) = p\mathcal{E}(x_3, x_4)$ and $Q(x_3, x_4) = q\mathcal{E}(x_3, x_4)$ where p, q, r are constants which satisfy $p^2 + q^2 \neq 0$. In this case,

$$a = 2px_1\mathcal{E}(x_3, x_4) + \xi(x_3, x_4), \qquad b = 2qx_2\mathcal{E}(x_3, x_4) + \eta(x_3, x_4),$$
$$c = (qx_1 + px_2)\mathcal{E}(x_3, x_4) + \gamma(x_3, x_4).$$

For the particular case of self-dual Walker manifolds one has the following:

Theorem 7.20. $\mathcal{C}_{a,b,c}$ is locally conformally Kaehler self-dual if and only if

$$a = x_1x_2\mathcal{D}(x_3, x_4) + x_1P(x_3, x_4) + x_2Q(x_3, x_4) + \xi(x_3, x_4),$$
$$b = -x_1x_2\mathcal{D}(x_3, x_4) + x_1S(x_3, x_4) + x_2T(x_3, x_4) + \eta(x_3, x_4),$$
$$c = \tfrac{1}{2}(x_2^2 - x_1^2)\mathcal{D}(x_3, x_4) - \tfrac{1}{2}x_1(Q(x_3, x_4) - T(x_3, x_4))$$
$$+\tfrac{1}{2}x_2(P(x_3, x_4) - S(x_3, x_4)) + \gamma(x_3, x_4),$$

where $Q_3 + T_3 = P_4 + S_4$.

Corollary 7.21. $\mathcal{C}_{a,b,c}$ is Kaehler self-dual if and only if

$$a = x_1x_2\mathcal{D}(x_3, x_4) + x_1P(x_3, x_4) + x_2Q(x_3, x_4) + \xi(x_3, x_4),$$
$$b = -x_1x_2\mathcal{D}(x_3, x_4) - x_1P(x_3, x_4) - x_2Q(x_3, x_4) + \eta(x_3, x_4),$$
$$c = \tfrac{1}{2}(x_2^2 - x_1^2)\mathcal{D}(x_3, x_4) - x_1Q(x_3, x_4) + x_2P(x_3, x_4) + \gamma(x_3, x_4).$$

Theorem 7.22. $\mathcal{C}_{a,b,c}$ is strictly locally conformally Kaehler and $\star$-Einstein if and only if the functions a, b and c have the forms

$$a = x_1 P(x_3, x_4) + x_2 Q(x_3, x_4) + \xi(x_3, x_4),$$
$$b = x_1 S(x_3, x_4) + x_2 T(x_3, x_4) + \eta(x_3, x_4),$$
$$c = -\tfrac{1}{2}x_1(Q(x_3, x_4) - T(x_3, x_4)) + \tfrac{1}{2}x_2(P(x_3, x_4) - S(x_3, x_4)) + \gamma(x_3, x_4),$$

where at least one of $P + S$ and $Q + T$ does not vanish everywhere, where $Q_3 + T_3 = P_4 + S_4$, and where $4(3S_3 - P_3 + 3Q_4 - T_4) = (P + S)^2 + (Q + T)^2$.

Theorem 7.23. $\mathcal{C}_{a,b,c}$ is strictly locally conformally Kaehler Einstein if and only if the functions a, b and c have the forms

$$a = x_1 P(x_3, x_4) + x_2 Q(x_3, x_4) + \xi(x_3, x_4),$$
$$b = x_1 S(x_3, x_4) + x_2 T(x_3, x_4) + \eta(x_3, x_4),$$
$$c = -\tfrac{1}{2}x_1(Q(x_3, x_4) - T(x_3, x_4)) + \tfrac{1}{2}x_2(P(x_3, x_4) - S(x_3, x_4)) + \gamma(x_3, x_4),$$

where at least one of the functions $P + S$ and $Q + T$ does not vanish everywhere and

$$8Q_4 - 4(P_3 - S_3) = 2Q(Q + T) + (P^2 - S^2),$$
$$8S_3 + 4(Q_4 - T_4) = 2S(P + S) - (Q^2 - T^2),$$
$$4(P_4 + S_4) = 4(Q_3 + T_3) = Q(P + 3S) - T(P - S).$$

Using Theorem 7.2, we easily obtain from Theorem 7.11 the following:

Corollary 7.24. $\mathcal{C}_{a,b,c}$ is Kaehler Einstein if and only if the functions a, b, c have one of the following forms where $\kappa \neq 0$ is constant:

1. *Form I* $(\tau = 8\kappa)$:

$$a = \kappa(x_1^2 - x_2^2) + x_1 P(x_3, x_4) + x_2 Q(x_3, x_4) + \xi(x_3, x_4),$$
$$b = \kappa(x_2^2 - x_1^2) - x_1 P(x_3, x_4) - x_2 Q(x_3, x_4) - \xi(x_3, x_4) + \tfrac{1}{\kappa}(P_3(x_3, x_4) - Q_4(x_3, x_4)),$$
$$c = 2\kappa x_1 x_2 - x_1 Q(x_3, x_4) + x_2 P(x_3, x_4) + \gamma(x_3, x_4).$$

2. *Form II* $(\tau = 0)$:

$$a = x_1 P(x_3, x_4) + x_2 Q(x_3, x_4) + \xi(x_3, x_4),$$
$$b = -x_1 P(x_3, x_4) - x_2 Q(x_3, x_4) + \eta(x_3, x_4),$$
$$c = -x_1 Q(x_3, x_4) + x_2 P(x_3, x_4) + \gamma(x_3, x_4),$$

where $P_3 = Q_4$.

Remark 7.25. In the case when the function c depends only on (x_3, x_4), $\mathcal{C}_{a,b,c}$ is Hermitian and Einstein if and only if the functions a and b have the forms

$$a = x_1 P(x_3, x_4) + x_2 Q(x_3, x_4) + \xi(x_3, x_4), \quad b = x_1 P(x_3, x_4) + x_2 Q(x_3, x_4) + \eta(x_3, x_4),$$

where $2P_3 = P^2$, $2Q_4 = Q^2$, and $P_4 + Q_3 = PQ$. We apply Lemma 2.13 to see the solution of this system is given by $P = -2a_4(a_0 + a_3 x_3 + a_4 x_4)^{-1}$ and $Q = -2a_3(a_0 + a_3 x_3 + a_4 x_4)^{-1}$. Note that if $P \cdot Q \neq 0$, then the structure $(g_{a,b,c}, J)$ is strictly locally conformally Kaehler, self-dual, Ricci flat and $\star$-Ricci flat, but the metric $g_{a,b,c}$ is not flat.

7.5   ALMOST KAEHLER WALKER FOUR-DIMENSIONAL MANIFOLDS

This section contains joint work with J. Davidov, J. C. Díaz-Ramos, Y. Matsushita, and O. Muškarov [93]. We adopt the notation of Example 7.1 to define $\mathcal{C}_{a,b,c}$. We refer to the discussion in Section 7.3 and to the survey [13]. Both ρ and $\rho^\star$ coincide in the Kaehler setting. However, in general $\rho^\star$ need not be a symmetric 2 tensor. We say that (M, g, J) is *weakly $\star$-Einstein* if $\rho^\star = \frac{\tau^\star}{m} g$ and that (M, g, J) is *$\star$-Einstein* if, in addition, $\tau^\star$ is constant.

Theorem 7.26. *$\mathcal{C}_{a,b,c}$ is almost Kaehler and self-dual if and only if*

$$a = x_1 x_2 \mathcal{D}(x_3, x_4) + x_1 P(x_3, x_4) + x_2 Q(x_3, x_4) + \xi(x_3, x_4),$$
$$b = -x_1 x_2 \mathcal{D}(x_3, x_4) - x_1 P(x_3, x_4) - x_2 Q(x_3, x_4) + \eta(x_3, x_4),$$
$$c = -\tfrac{1}{2} x_1^2 \mathcal{D}(x_3, x_4) + \tfrac{1}{2} x_2^2 \mathcal{D}(x_3, x_4) + x_1 U(x_3, x_4) + x_2 V(x_3, x_4) + \gamma(x_3, x_4).$$

Theorem 7.27. *$\mathcal{C}_{a,b,c}$ is almost Kaehler and $\star$-Einstein if and only if*

$$a = \kappa(x_1^2 - x_2^2) + x_1 P(x_3, x_4) + x_2 Q(x_3, x_4) + \xi(x_3, x_4),$$
$$b = \kappa(x_2^2 - x_1^2) - x_1 P(x_3, x_4) - x_2 Q(x_3, x_4) + \eta(x_3, x_4),$$
$$c = 2\kappa x_1 x_2 + x_1 U(x_3, x_4) + x_2 V(x_3, x_4) + \gamma(x_3, x_4),$$

where κ is a constant and $2(P_3 + V_3 - Q_4 + U_4) = (P - V)^2 + (Q + U)^2 + 4\kappa(\xi + \eta)$. In this case, the scalar and $\star$-scalar curvatures are constant, $\tau = \tau^\star = 8\kappa$.

Corollary 7.28.

1. *$\mathcal{C}_{a,b,c}$ is almost Kaehler, self-dual and $\star$-Einstein if and only if $\kappa = 0$ in Theorem 7.27.*

2. *If $a = a(x_3, x_4)$ (resp. $b = b(x_3, x_4)$), then $\mathcal{C}_{a,b,c}$ is almost Kaehler and $\star$-Einstein if and only if the function $b = b(x_3, x_4)$ (resp. $a = a(x_3, x_4)$) and*

$c = x_1 U(x_3, x_4) + x_2 V(x_3, x_4) + \gamma(x_3, x_4)$, *where* $2(V_3 + U_4) = V^2 + U^2$.

3. *If* $c = c(x_3, x_4)$, *then* $\mathcal{C}_{a,b,c}$ *is almost Kaehler and* $\star$*-Einstein if and only if*

$$a = x_1 P(x_3, x_4) + x_2 Q(x_3, x_4) + \xi(x_3, x_4),$$

$$b = -x_1 P(x_3, x_4) - x_2 Q(x_3, x_4) + \eta(x_3, x_4),\ \text{where}\ 2(P_3 - Q_4) = P^2 + Q^2.$$

The Einstein equations for $\mathcal{M}_{a,b,c}$ are given in Section 5.3.

Theorem 7.29. $\mathcal{C}_{a,b,c}$ *is strictly almost Kaehler Einstein if and only if*

$$a = x_1 P(x_3, x_4) + x_2 Q(x_3, x_4) + \xi(x_3, x_4),$$
$$b = -x_1 P(x_3, x_4) - x_2 Q(x_3, x_4) + \eta(x_3, x_4),$$
$$c = x_1 U(x_3, x_4) + x_2 V(x_3, x_4) + \gamma(x_3, x_4),$$

where $(V - P)^2 + (U + Q)^2 \neq 0$ *and*

$$2(V_3 - Q_4) = V^2 - VP + Q^2 + UQ, \quad 2(P_3 + U_4) = P^2 - VP + U^2 + UQ,$$
$$Q_3 + U_3 - P_4 + V_4 = PQ + UV.$$

Corollary 7.30.

1. *If* $\mathcal{C}_{a,b,c}$ *is strictly almost Kaehler Einstein,* $\mathcal{C}_{a,b,c}$ *is self-dual, Ricci flat and* $\star$*-Ricci flat.*

2. *If* $c = c(x_3, x_4)$, $\mathcal{C}_{a,b,c}$ *is strictly almost Kaehler Einstein if and only if*

$$a = x_1 P(x_3, x_4) + x_2 Q(x_3, x_4) + \xi(x_3, x_4),$$
$$b = -x_1 P(x_3, x_4) - x_2 Q(x_3, x_4) + \eta(x_3, x_4),$$

where $P^2 + Q^2 \neq 0, 2P_3 = P^2, 2Q_4 = -Q^2$, *and* $Q_3 - P_4 = PQ$.

Remark 7.31. Near any point where $P^2 + Q^2 \neq 0$, we may use Lemma 2.13 (after first replacing Q by $-Q$) to see the solution of the system described in Corollary 7.30 (2) is given by setting $P = -2p(px_3 + qx_4 + r)^{-1}$ and $Q = 2q(px_3 + qx_4 + r)^{-1}$ where p, q, r are constants and where $p^2 + q^2 \neq 0$.

Corollary 7.32. *If the function $a = a(x_3, x_4)$ (resp. $b = b(x_3, x_4)$), then $C_{a,b,c}$ is strictly almost Kaehler Einstein if and only if the function $b = b(x_3, x_4)$ (resp. $a = a(x_3, x_4)$) and*

$$c = x_1 U(x_3, x_4) + x_2 V(x_3, x_4) + \gamma(x_3, x_4),$$

where $U^2 + V^2 \neq 0, 2U_4 = U^2, 2V_3 = V^2$, and $U_3 + V_4 = UV$.

Remark 7.33. Near any point where $U^2 + V^2 \neq 0$, we may use a variant of Lemma 2.13 to see that the solution of the system in Corollary 7.32 is given by $U = -2p(qx_3 + px_4 + r)^{-1}$ and $V = -2q(qx_3 + px_4 + r)^{-1}$ where p, q, r are constants and $p^2 + q^2 \neq 0$.

It is well-known [16, 211, 212] that in the definite case there are no strictly almost Kaehler structures of constant sectional curvature. In the indefinite case, however, one can construct such structures.

Theorem 7.34. *If $c = 0$, $C_{a,b,c}$ is strictly almost Kaehler and of constant sectional curvature if and only if the functions a and b have the form*

$$a = x_1 P(x_3, x_4) + x_2 Q(x_3, x_4) + \xi(x_3, x_4),$$
$$b = -x_1 P(x_3, x_4) - x_2 Q(x_3, x_4) + \eta(x_3, x_4),$$

where $P^2 + Q^2 \neq 0$ and where

$$2P_3 = P^2, \quad 2P_4 = -PQ, \quad 2Q_3 = PQ, \quad 2Q_4 = -Q^2,$$
$$2\xi_{44} + 2\eta_{33} + (\xi_3 + \eta_3)P - (\xi_4 + \eta_4)Q + \xi P^2 + \eta Q^2 = 0.$$

In particular, every such a structure $(g_{a,b,c}, J)$ is flat.

One also has that

Theorem 7.35. *If $a = b = 0$, $C_{a,b,c}$ is of constant sectional curvature if and only if*

$$c = x_1 U(x_3, x_4) + x_2 V(x_3, x_4) + \gamma(x_3, x_4),$$

where $U^2 + V^2 \neq 0$ and where U, V, and γ satisfy $2U_3 = UV, 2U_4 = U^2, 2V_3 = V^2, 2V_4 = UV$, and $\gamma_{34} = \gamma U_3$. In particular, every such a structure $(g_{a,b,c}, J)$ is flat.

Example. Let p, q, r be arbitrary constants and $p^2 + q^2 \neq 0$. Set

$$a = (-2px_1 + 2qx_2)(px_3 + qx_4 + r)^{-1}, \quad b = (2px_1 - 2qx_2)(px_3 + qx_4 + r)^{-1}, \quad c = 0,$$

or

$$a = 0, \quad b = 0, \quad c = -(2px_1 + 2qx_2)(qx_3 + px_4 + r)^{-1}.$$

Then it follows from Theorems 7.34 and 7.35, and Remarks 7.31 and 7.33 (see Lemma 7.36 below) that the proper almost Hermitian structure $(g_{a,b,c}, J)$ determined by means of the functions a, b, c defined above is strictly almost Kaehler and the metric g is flat.

Lemma 7.36.   *If $a = b = 0$, then $\mathcal{C}_{a,b,c}$ is an almost Kaehler manifold which is Kaehler if and only if $c = c(x_3, x_4)$.*

CHAPTER 8

Special Walker Manifolds

8.1   INTRODUCTION

We adopt the notation of Example 5.2. In this Chapter, we consider Walker manifolds $\mathcal{M}_{a,b,c}$ where two of the defining functions vanish. This permits a somewhat more detailed analysis than is available in the general context. On the other hand, this is not an invariant notion.

In Section 8.3, we set $a = b = 0$ and examine when the manifolds $\mathcal{M}_{0,0,c}$ are Osserman, self-dual, or anti-self-dual. We determine the eigenvalues of the Weyl operator in the self-dual and in the anti-self-dual contexts and study the commutativity conditions of Definition 1.27 for this family. We give examples of conformally Osserman manifolds whose eigenvalue structures exhibit all the possible algebraic types of Theorem 1.18; certain of these manifolds are strictly conformally Osserman, i.e., they are conformally Osserman but are not in the conformal class of any Osserman metric. We examine questions of geodesic completeness and Ricci blowup as well as study curvature homogeneity.

In Section 8.4, we change our focus slightly and set $b = c = 0$. We examine questions of curvature homogeneity for the family

$$\mathcal{M}_f := \mathcal{M}_{a,0,0} \quad \text{where} \quad a = -2f(x_4) \in C^{\infty}(\mathbb{R}) .$$

We will show that if $f^{(2)} > 0$, then $\mathcal{M}_f$ is curvature homogeneous and is curvature modeled on a symmetric space. If additionally $f^{(3)} > 0$, then $\mathcal{M}_f$ is 1 curvature homogeneous and 1 curvature modeled on a homogeneous space; $\mathcal{M}_f$ is homogeneous if and only if one has that $f^{(2)}(x_4) = \alpha e^{\lambda x_4}$, for some $\alpha, \lambda \in \mathbb{R}$.

8.2   HISTORY

The study of commutativity properties of natural operators associated with the curvature tensor has received much attention in recent years. We refer to [48] for an overview of the field and only mention a few high points here.

The field could be said to begin with the seminal work of Stanilov and Videv [246] where commutation questions were first introduced, although Ricci semi-symmetric manifolds had been studied previously in the literature. Tsankov [253] subsequently classified all the hypersurfaces in flat Euclidean space which were curvature–curvature commuting. Subsequent to that time, a number of authors have contributed to the field. Originally, the Riemannian setting has been studied [54, 55, 158, 174] but subsequently the pseudo-Riemannian setting was a fertile field of investigation as well; we refer to [53, 58] for further details.

As noted previously, curvature homogeneity also has been a fertile field of investigation. Again, the Riemannian setting was the original context [232, 235, 236, 240, 249]. Three dimensional Lorentzian curvature homogeneous manifolds have been studied extensively [64, 65, 72, 73, 74, 232] but the four dimensional setting has also been important both in the Lorentzian [202] and the (2, 2) signature settings [99, 114]. The higher dimensional context has been examined as well [113, 146, 147, 149]. We refer to [140] for a more extensive bibliography than it is possible to present here and also to the seminal work of [226].

8.3   CURVATURE COMMUTING CONDITIONS

In this section, we report on work of [50, 51, 52]. We set $a = b = 0$ and consider the manifold $\mathcal{M}_{0,0,c}$ in the notation of Example 5.2. There are two separate families which deserve special attention:

$$c = x_1 P(x_3, x_4) + x_2 Q(x_3, x_4) + \gamma(x_3, x_4) \text{ where } P^2 = 2P_4,$$
$$Q^2 = 2Q_3, \quad PQ = P_3 + Q_4, \tag{8.1}$$

and

$$c = x_1 P(x_3, x_4) + x_2 Q(x_3, x_4) + \gamma(x_3, x_4) \text{ where } P_3 = Q_4 . \tag{8.2}$$

We note by Lemma 2.13 that the conditions on P and Q in Equation (8.1) are equivalent to the conditions

$$P^2 = 2P_4, \quad Q^2 = 2Q_3, \quad P_3 = Q_4 = \tfrac{1}{2} PQ .$$

Theorems 5.5, 5.6, and 5.8 imply:

Lemma 8.1.   *The curvature and Weyl curvature tensor of $\mathcal{M}_{0,0,c}$ take the form:*

$$R_{1334} = -\tfrac{1}{4}(c_1 c_2 - 2c_{13}), \quad R_{1314} = \tfrac{1}{2}c_{11}, \quad R_{2324} = \tfrac{1}{2}c_{22},$$

$$R_{1434} = -\tfrac{1}{4}(-c_1^2 + 2c_{14}), \quad R_{1324} = \tfrac{1}{2}c_{12}, \quad R_{1423} = \tfrac{1}{2}c_{12},$$

$$R_{2334} = -\tfrac{1}{4}(c_2^2 - 2c_{23}), \quad R_{2434} = -\tfrac{1}{4}(-c_1 c_2 + 2c_{24}),$$

$$R_{3434} = -\tfrac{1}{2}(-cc_1 c_2 + 2c_{34}),$$

$$W_{11}^- = -\tfrac{1}{3}c_{12}, \quad W_{12}^- = \tfrac{1}{4}(c_{11} + c_{22}), \quad W_{13}^- = 0,$$

$$W_{22}^- = -\tfrac{2}{3}c_{12}, \quad W_{23}^- = \tfrac{1}{4}(c_{11} - c_{22}), \quad W_{33}^- = \tfrac{1}{3}c_{12},$$

$$W_{11}^+ = \tfrac{1}{6}c_{12} + 2cc_{13} - 2c_{34}, \quad W_{22}^+ = \tfrac{1}{3}c_{12}, \quad W_{33}^+ = W_{11}^+ - \tfrac{1}{3}c_{12},$$

$$W_{13}^+ = W_{11}^+ - \tfrac{1}{6}c_{12}, \quad W_{23}^+ = W_{12}^+ = \tfrac{1}{2}(cc_{12} + c_{13} - c_{24}).$$

One has the following results for this family:

Theorem 8.2.

1. $\mathcal{M}_{0,0,c}$ *is self-dual if and only if* $c = x_1 P(x_3, x_4) + x_2 Q(x_3, x_4) + \gamma(x_3, x_4)$.

2. $\mathcal{M}_{0,0,c}$ *is anti-self-dual if and only if*

$$c = x_1 P(x_3, x_4) + x_2 Q(x_3, x_4) + \gamma(x_3, x_4) + \xi(x_1, x_4) + \eta(x_2, x_3)$$

   with $P_3 = Q_4$ *and* $c P_3 - x_1 P_{34} - x_2 P_{33} - \gamma_{34} = 0$.

3. *The following assertions are equivalent:*

   (a) $\mathcal{M}_{0,0,c}$ *is Einstein.*

   (b) c *is as in* Equation (8.1).

   (c) $\rho = 0$.

   (d) $\mathcal{M}_{0,0,c}$ *is Osserman.*

   (e) *The only (possibly) non-zero curvature is* $R_{3434} = \frac{1}{2} c c_1 c_2 - c_{34}$.

It is a crucial feature of these examples that Ricci flat, Einstein, and Osserman are equivalent conditions; this is not the case, of course, for general Walker manifolds of signature $(2, 2)$. Furthermore, if $\mathcal{M}_{0,0,c}$ is Einstein, then c is affine in (x_1, x_2) so $\mathcal{M}_{0,0,c}$ is a twisted Riemannian extension as discussed in Section 3.5.1. This will be a consistent theme in what follows. We refer to Theorem 8.6 below for other conditions which are equivalent to the Einstein condition.

Theorem 8.2 and Lemma 2.13 yield the following:

Corollary 8.3. *Suppose that $\mathcal{M}_{0,0,c}$ is Osserman. Then:*

1. $\mathcal{M}_{0,0,c}$ *is a twisted Riemannian extension* $g_{D,\phi}$ *given by* Equation (3.4).

2. c *is globally defined if and only if* $c = c(x_3, x_4)$.

3. *If* $R_{3434} = 0$, *then* $\mathcal{M}_{0,0,c}$ *is flat.*

4. *If* $R_{3434} \neq 0$, *then* $\mathcal{M}_{0,0,c}$ *is Osserman and 2-step nilpotent Ivanov–Petrova.*

This result can be used to construct examples of Osserman and Ivanov–Petrova manifolds where the Jordan normal form varies from point to point. Let $p_\lambda^\pm$ (resp. $m_\lambda^\pm$) be the characteristic polynomial (resp. minimal polynomial) of the Weyl curvature operator $W^\pm$ at a point $P \in \mathbb{R}^4$.

Theorem 8.4.

1. *If* $\mathcal{M}_{0,0,c}$ *is self-dual, then* $p_\lambda^+ = -\lambda^3$ *and*

   (a) $m_\lambda^+ = \lambda^3$ *if* $c_{13} - c_{24} \neq 0$.

(b) $m_\lambda^+ = \lambda^2$ if $c_{13} - c_{24} = 0$ and $cc_{13} - c_{34} \neq 0$.

(c) $m_\lambda^+ = \lambda$ if $c_{13} - c_{24} = 0$ and $cc_{13} - c_{34} = 0$.

2. If $\mathcal{M}_{0,0,c}$ is anti-self-dual, then $\mathrm{Spec}\{W^-\} = \{0, \pm\frac{1}{2}(-c_{11}c_{22})^{\frac{1}{2}}\}$. Also

(a) $m_\lambda^- = \lambda(\lambda^2 + \frac{1}{4}c_{11}c_{22})$ if $c_{11}c_{22} \neq 0$.

(b) $m_\lambda^- = \lambda^3$ if either $c_{11} = 0$ or $c_{22} = 0$ but not both.

(c) $m_\lambda^- = \lambda$ if $c_{11} = c_{22} = 0$.

We now turn our attention to commutativity conditions and recall the notation established in Definition 1.27. We have:

Theorem 8.5. *The following assertions are equivalent:*

1. $\mathcal{M}_{0,0,c}$ *is Jacobi–Ricci commuting.*

2. $\mathcal{M}_{0,0,c}$ *is curvature–Ricci commuting.*

3. $\mathcal{M}_{0,0,c}$ *is curvature–curvature commuting.*

4. c *satisfies the conditions of* Equation (8.2).

We study when $\mathcal{M}_{0,0,c}$ is Jacobi–Jacobi commuting; Equation (8.1) appears once again.

Theorem 8.6. *The following assertions are equivalent:*

1. $\mathcal{M}_{0,0,c}$ *is curvature–Jacobi commuting.*

2. $\mathcal{M}_{0,0,c}$ *is Jacobi–Jacobi commuting.*

3. $\mathcal{M}_{0,0,c}$ *is Osserman.*

4. c *satisfies* Equation (8.1).

Lemma 2.13 shows that the conditions of Theorem 8.2 (3) are very rigid. On the other hand, the condition of Theorem 8.5 (4) that $P_3 = Q_4$ is, of course, nothing but the condition that

$$\omega := P \, dx_4 + Q \, dx_3$$

is a closed 1 form. Thus, there are many examples.

We continue the discussion of Section 6.5.3. Let D be a torsion free connection on a 2 dimensional manifold. The conditions given in Equations (8.1) and (8.2) can be expressed in terms of a twisted Riemannian extension as follows.

Lemma 8.7. *Let*

$$c = x_1 P(x_3, x_4) + x_2 Q(x_3, x_4) + \gamma(x_3, x_4) \ .$$

Then, taking coordinates (x_3, x_4) on the affine manifold, $\mathcal{M}_{0,0,c}$ is a twisted Riemannian extension $g_{D,\phi}$ given by Equation (3.4), *where*

$$\phi_{33} = \phi_{44} = \Gamma_{33}{}^3 = \Gamma_{33}{}^4 = \Gamma_{44}{}^3 = \Gamma_{44}{}^4 = 0,$$
$$\phi_{34} = \gamma, \quad \Gamma_{34}{}^3 = -\tfrac{1}{2}P, \quad \Gamma_{34}{}^4 = -\tfrac{1}{2}Q \ .$$

Moreover,

1. *$\rho_a^D = 0$ if and only if c satisfies* Equation (8.2).

2. *$\rho_s^D = 0$ if and only if $\rho^D = 0$ if and only if c satisfies* Equation (8.1).

If $a = b = 0$, then $\rho_s^D = 0 \Leftrightarrow \rho^D = 0$. This is, of course, a reflection of the equivalence of conditions (1) and (2) in Lemma 2.13. However, this is not the case for a more general affine extension. In contrast to the situation we have been considering where there are essentially two different conditions on c, there are four cases of interest for a general twisted Riemannian extension (cf. Theorem 6.28).

If $\mathcal{M}_{0,0,c}$ is conformally Osserman, let Spec_W denote the set of eigenvalues of $\mathcal{J}_W$ where each eigenvalue is repeated according to multiplicity and let m_λ denote the minimal polynomial of $\mathcal{J}_W$. We have the following examples which illustrate different phenomena of conformally Osserman manifolds:

Theorem 8.8. *The following manifolds $\mathcal{M}_{0,0,c}$ are conformally Osserman.*

1. *The Jordan normal form does not change from point to point:*

    (a) *$c = x_1^2 - x_2^2 \Rightarrow m_\lambda = \lambda(\lambda^2 - \tfrac{1}{4})$ and $\text{Spec}_W = \{0, 0, \pm\tfrac{1}{2}\}$.*

    (b) *$c = x_1^2 + x_2^2 \Rightarrow m_\lambda = \lambda(\lambda^2 + \tfrac{1}{4})$ and $\text{Spec}_W = \{0, 0, \pm\tfrac{\sqrt{-1}}{2}\}$.*

(c) $c = x_1 x_4 + x_3 x_4 \Rightarrow m_\lambda = \lambda^2$ and $\mathrm{Spec}_W = \{0\}$.

(d) $c = x_1^2 \Rightarrow m_\lambda = \lambda^3$ and $\mathrm{Spec}_W = \{0\}$.

2. $\mathrm{Spec}_W = \{0\}$ but the Jordan normal form changes from point to point:

(a) $c = x_2 x_4^2 + x_3^2 x_4 \Rightarrow m_\lambda = \begin{cases} \lambda^3 & \text{if } x_4 \neq 0, \\ \lambda^2 & \text{if } x_4 = 0, \ x_3 \neq 0, \\ \lambda & \text{if } x_3 = x_4 = 0. \end{cases}$

(b) $c = x_2 x_4^2 + x_3 x_4 \Rightarrow m_\lambda = \begin{cases} \lambda^3 & \text{if } x_4 \neq 0, \\ \lambda^2 & \text{if } x_4 = 0. \end{cases}$

(c) $c = x_1 x_3^2 \Rightarrow m_\lambda = \begin{cases} \lambda^3 & \text{if } x_3 \neq 0, \\ \lambda & \text{if } x_3 = 0. \end{cases}$

(d) $c = x_1 x_3 + x_2 x_4 \Rightarrow m_\lambda = \begin{cases} \lambda^2 & \text{if } x_1 x_3 + x_2 x_4 \neq 0, \\ \lambda & \text{if } x_1 x_3 + x_2 x_4 = 0. \end{cases}$

3. The eigenvalues can change from point to point:

(a) $c = x_1^4 + x_1^2 - x_2^4 - x_2^2 \Rightarrow \mathrm{Spec}_W = \{0, 0, \pm\frac{1}{2}\sqrt{(6x_1^2 + 1)(6x_2^2 + 1)}\}$.

(b) $c = x_1^4 + x_1^2 + x_2^4 + x_2^2 \Rightarrow \mathrm{Spec}_W = \{0, 0, \pm\frac{1}{2}\sqrt{-(6x_1^2 + 1)(6x_2^2 + 1)}\}$.

(c) $c = x_1^3 - x_2^3 \Rightarrow \mathrm{Spec}_W = \{0, 0, \pm\frac{3}{2}\sqrt{x_1 x_2}\}$.

We note that (1a) has Type Ia, that (1b) has Type Ib, that (1c) has Type II, and that (1d) has Type III in the classification of Theorem 1.18. Thus, all algebraic types can be realized geometrically in this context.

Recall that $\mathcal{M}$ is said to be *strictly conformally Osserman* if it is conformally Osserman but it does not belong to the conformal class of any Osserman metric; we refer to related work by Nikolayevsky [210] in the Riemannian setting.

Theorem 8.9.   *Let $\mathcal{M}_{0,0,c}$ be as in Theorem 8.8.*

1. *The manifolds of (1a), (1b), and (1d) are strictly conformally Osserman.*

2. *Only (1c) defines a geodesically complete manifold; the remaining manifolds exhibit Ricci blowup and therefore can not be embedded isometrically in a geodesically complete manifold.*

3. *The manifolds other than (1d) are not curvature homogeneous.*

4. The manifold (1d) is curvature homogeneous, not 1 *curvature homogeneous, and the group of isometries acts transitively on* $\mathcal{O} := \mathbb{R}^4 - \{(0, x_2, x_3, x_4)\}$.

Remark 8.10.   Let R_Λ be the induced action of the curvature on $\Lambda^2(T^*M)$. We note that Derdzinski [98] showed a 4 dimensional Riemannian manifold is curvature homogeneous if and only if R_Λ has constant eigenvalues; furthermore if such a manifold is Einstein, then it is locally symmetric. In (2d), $\mathrm{Spec}\{R_\Lambda\} = \{0\}$ but the manifold is not curvature homogeneous. Thus, this result of Derdzinski fails in signature $(2, 2)$; we refer to Derdzinski [99] for additional results in this direction.

Remark 8.11.   In Case (1) above, the group of isometries acts transitively on the proper open subset $\mathcal{O}$ of the manifold

$$\mathcal{M}_{0,0,x_1^2} \ .$$

Thus, $\mathcal{O}$ is a homogeneous space which is geodesically incomplete; this can not happen in the Riemannian setting.

8.4   CURVATURE HOMOGENEOUS STRICT WALKER MANIFOLDS

This reports on work with C. Dunn [114] and generalizes previous work by Derdzinski [98] in the area. We refer to [140] for other examples. Let $f = f(x_4)$. We set

$$a = -2f, \quad b = c = 0$$

to define, using the notation of Example 5.2,

$$\mathcal{M}_f := \mathcal{M}_{a,0,0} \ .$$

This strict Walker manifold is a twisted Riemannian extension where the connection on $\mathbb{R}^2$ is flat as was discussed in Section 3.5.2. By Theorem 3.17, all manifolds in this family are nilpotent Ivanov–Petrova. Similarly, by Theorem 3.16, any such manifold is nilpotent Osserman. It will follow from Theorem 8.12 and Lemma 2.12 that $\mathcal{M}_f$ is symmetric if and only if $f^{(3)} = 0$. Furthermore, if $\mathcal{M}_f$ is symmetric, then either $\mathcal{M}_f$ is flat, or $\mathcal{M}_f$ is locally isometric to one of the following two examples:

$$\mathcal{M}_{x_4^2} \quad \text{or} \quad \mathcal{M}_{-x_4^2} \ .$$

It turns out that the $\mathrm{sign}(f^{(2)})$ is a local isometry invariant; we shall therefore usually assume $f^{(2)} > 0$ as the case $f^{(2)} < 0$ can be handled similarly.

The following is an immediate consequence of Theorem 3.9:

Theorem 8.12.   *The manifolds* $\mathcal{M}_f$ *satisfy:*

1. $\nabla^k R(\partial_{x_3}, \partial_{x_4}, \partial_{x_4}, \partial_{x_3}; \partial_{x_4}, \dots, \partial_{x_4}) = f^{(k+2)}$; *the remaining components vanish.*

2. *All geodesics extend for infinite time.*

3. *If* $P \in \mathbb{R}^4$, *then* $\exp_P : T_P \mathbb{R}^4 \to \mathbb{R}^4$ *is a diffeomorphism.*

4. *All the local Weyl scalar invariants of* $\mathcal{M}_f$ *vanish so* $\mathcal{M}_f$ *is VSI.*

For $p \geq 2$ and $f^{(3)} > 0$, set

$$\alpha_p(f) := f^{(p+2)} \{f^{(2)}\}^{p-1} \{f^{(3)}\}^{-p} .$$

We have the following curvature homogeneity results:

Theorem 8.13.   *Assume* $f^{(2)} > 0$.

1. $\mathcal{M}_f$ *is curvature modeled on the symmetric space* $\mathcal{M}_{x_4^2}$.

2. *Assume additionally that* $f^{(3)} > 0$.

   (a) $\mathcal{M}_f$ *is* 1 *curvature modeled on the homogeneous space* $\mathcal{M}_{e^{x_4}}$.

   (b) $\mathcal{M}_f$ *is homogeneous if and only if* $f^{(2)} = \alpha e^{\lambda x_4}$ *for some* $\alpha, \lambda \in \mathbb{R}$.

Theorem 8.14.   *Assume that* $f_i^{(2)} > 0$ *and* $f_i^{(3)} > 0$.

1. *Let* f_i *be real analytic. Assume* $\alpha_p(f_1)(P_1) = \alpha_p(f_2)(P_2)$ *for all* $p \geq 2$. *Then there exists a real analytic isometry*

$$\phi : (\mathcal{M}_{f_1}, P_1) \to (\mathcal{M}_{f_2}, P_2) .$$

2. *If there is an isomorphism* Φ *from the* k *curvature model* $\mathfrak{M}^k(\mathcal{M}_{f_1}, P_1)$ *to the* k *curvature model* $\mathfrak{M}^k(\mathcal{M}_{f_2}, P_2)$, *then*

$$\alpha_p(f_1)(P_1) = \alpha_p(f_2)(P_2) \quad for \quad 2 \leq p \leq k .$$

We work in the real analytic context henceforth so Theorem 8.14 applies. The Walker manifold $\mathcal{M}_f$ can be realized as a hypersurface. Give $\mathbb{R}^{2,3} := \mathrm{Span}\{e_1, e_2, \tilde{e}_1, \tilde{e}_2, \check{e}\}$ a signature $(2,3)$ inner-product:

$$\langle e_i, \tilde{e}_i \rangle = 1, \quad \langle \check{e}, \check{e} \rangle = 1 .$$

Let $\mathcal{H}_f$ be the hypersurface defined by the embedding:

$$\Psi(x_1, x_2, x_3, x_4) := x_1 e_1 + x_2 e_2 + x_3 \tilde{e}_1 + x_4 \tilde{e}_2 + (\tfrac{1}{2} x_1^2 + f(x_4)) \check{e} .$$

Theorem 8.15.   *If* f *is real analytic, then* $(\mathcal{M}_f, P)$ *is isometric to* $(\mathcal{H}_f, P)$.

Let $\mathcal{G}$ be the Lie group of isometries of $\mathcal{M}_f$ and let $\mathfrak{g}$ be the associated Lie algebra.

Theorem 8.16.   *Let f be real analytic.*

1. *If $f^{(2)} = 0$, then* $\dim \mathfrak{g} = 10$.

2. *If $f^{(2)} = k \neq 0$, then* $\dim \mathfrak{g} = 8$.

3. *If $f^{(2)} = \alpha e^{\lambda x_4}$ for $\alpha \neq 0$ and $\lambda \neq 0$, then* $\dim \mathfrak{g} = 6$.

4. *If $f^{(2)} \neq \alpha e^{\lambda x_4}$ and if $f^{(2)} \neq \alpha(x_4 + \beta)^n$ for $n \in \mathbb{N}$, then* $\dim \mathfrak{g} = 5$.

Bibliography

[1] B. E. Abdalla, and F. Dillen, A Ricci-semi-symmetric hypersurface of Euclidean space which is not semi-symmetric, *Proc. Amer. Math. Soc.* **130** (2002), 1805–1808. DOI: 10.1090/S0002-9939-01-06220-7

[2] R. Abounasr, A. Belhaj, J. Rasmussen, and E. H. Saidi, Superstring theory on pp waves with *ADE* geometries, *J. Phys. A* **39** (2006), 2797–2841. DOI: 10.1088/0305-4470/39/11/015

[3] M. Abramowicz, O. M. Blaes, J. Horák, W. Kluźniak, and P. Rebusco, Epicyclic oscillations of fluid bodies. II. Strong gravity, *Classical Quantum Gravity* **23** (2006), 1689–1696. DOI: 10.1088/0264-9381/23/5/014

[4] M. Abramowicz, and W. Kluźniak, Epicyclic orbital oscillations in Newton's and Einstein's dynamics, *Gen. Relativity Gravitation* **35** (2003), 69–77. DOI: 10.1023/A:1021354928292

[5] J. Adams, Vector fields on spheres, *Ann. of Math.* **75** (1962), 603–632. DOI: 10.2307/1970213

[6] W. Adkins, and S. Weintraub, Algebra - an approach via module theory, *Graduate Texts in Mathematics* **136**, Springer-Verlag, New York (1992).

[7] Z. Afifi, Riemann extensions of affine connected spaces, *Quart. J. Math., Oxford Ser. (2)* **5** (1954), 312–320. DOI: 10.1093/qmath/5.1.312

[8] D. Alekseevsky, N. Blažić, N. Bokan, and Z. Rakić, Self-duality and pointwise Osserman spaces, *Arch. Math. (Brno)* **35** (1999), 193–201.

[9] D. Alekseevsky, N. Blažić, V. Cortés, and S. Vukmirović, A class of Osserman spaces, *J. Geom. Phys.* **53** (2005), 345–353. DOI: 10.1016/j.geomphys.2004.07.004

[10] D. Alekseevsky, V. Cortés, A. Galaev, and T. Leistner, Cones over pseudo-Riemannian manifolds and their holonomy. arxiv.org/abs/0707.3063

[11] W. Ambrose, and I. M. Singer, On homogeneous Riemannian manifolds, *Duke Math. J.* **2** (1958), 647–669. DOI: 10.1215/S0012-7094-58-02560-2

[12] V. Apostolov, Generalized Goldberg-Sacks theorems for pseudo Riemannian four-manifolds, *J. Geom. Phys.* **27** (1998), 185–198. DOI: 10.1016/S0393-0440(97)00075-2

[13] V. Apostolov, and T. Draghici, The curvature and the integrability of almost Kaehler manifolds: a survey, *Symplectic and contact topology: interactions and perspectives.* Toronto, ON/Montreal, QC, 2001, 25–53, Fields Inst. Commun. **35** Amer. Math. Soc., Providence, RI, 2003.

[14] V. Apostolov, and P. Gauduchon, The Riemannian Goldberg-Sacks theorems, *Internat. J. Math.* **8** (1997), 421–439. DOI: 10.1142/S0129167X97000214

[15] T. Arias-Marco, and O. Kowalski, Classification of locally homogeneous affine connections with arbitrary torsion on 2-dimensional manifolds, *Monatsh. Math* **153** (2008), 1–18. DOI: 10.1007/s00605-007-0494-0

[16] J. Armstrong, An ansatz for almost-Kaehler, Einstein 4-manifolds, *J. Reine Angew. Math.* **542** (2002), 53–84.

[17] M. F. Atiyah, R. Bott, and A. Shapiro, Clifford Modules, *Topology* **3** suppl. 1 (1964), 3–38. DOI: 10.1016/0040-9383(64)90003-5

[18] M. Barros, A. Romero, J. L. Cabrerizo, and M. Fernández, The Gauss-Landau-Hall problem on Riemannian surfaces, *J. Math. Phys.* **46** (2005), 112905 15 pp. DOI: 10.1063/1.2136215

[19] W. Batat, G. Calvaruso, and B. De Leo, Homogeneous structures on Lorentzian three-manifolds admitting a parallel null vector field, preprint.

[20] A. Bejancu, and H. R. Faran, Foliations and Geometric Structures, *Mathematics and Its Applications* (Springer), **580**, Springer, Dordrecht, 2006.

[21] M. Belger, and O. Kowalski, Riemannian metrics with the prescribed curvature tensor and all its covariant derivatives at one point, *Math. Nachr.* **168** (1994), 209–225.

[22] S. Bellucci, A. V. Glajinsky, and E. Latini, Making the hyper-Kähler structure of $N = 2$ quantum string manifest, *Phys. Rev. D* **70** (2004), 024013 5 pp. DOI: 10.1103/PhysRevD.70.024013

[23] L. Bérard Bergery, and A. Ikemakhen, Sur l'holonomie des variétés pseudo riemanniennes de signature (n, n), *Bull. Soc. Math. France* **125** (1997), 93–114.

[24] E. Bergshoeff, and E. Sezgin, Self-dual supergravity theories in $2 + 2$ dimensions, *Phys. Lett. B* **292** (1992), 87–92. DOI: 10.1016/0370-2693(92)90612-8

[25] J. Berndt, Three-dimensional Einstein-like manifolds, *Differential Geom. Appl.* **2** (1992), 385–397. DOI: 10.1016/0926-2245(92)90004-7

[26] A. Besse, Einstein manifolds, *Springer-Verlag*, Berlin 1986.

[27] R. L. Bishop, and S. I. Goldberg, On conformally flat spaces with commuting curvature and Ricci transformations, *Canad. J. Math.* **24** (1972), 799–804.

[28] D. Blair, J. Davidov, and O. Muškarov, Isotropic Kaehler hyperbolic twistor spaces, *J. Geom. Phys.* **52** (2004), 74–88. DOI: 10.1016/j.geomphys.2004.02.002

[29] N. Blažić, Natural curvature operators of bounded spectrum, *Differential Geom. Appl.* **24** (2006), 563–566. DOI: 10.1016/j.difgeo.2006.02.004

[30] N. Blažić, N. Bokan, and P. Gilkey, A note on Osserman Lorentzian manifolds, *Bull. London Math. Soc.* **29** (1997), 227–230. DOI: 10.1112/S0024609396002238

[31] N. Blažić, N. Bokan, and Z. Rakić, A note on the Osserman conjecture and isotropic covariant derivative of curvature, *Proc. Amer. Math. Soc.* **128** (2000), 245–253. DOI: 10.1090/S0002-9939-99-05131-X

[32] N. Blažić, N. Bokan, and Z. Rakić, Osserman pseudo-Riemannian manifolds of signature (2, 2), *J. Aust. Math. Soc.* **71** (2001), 367–395. DOI: 10.1017/S1446788700003001

[33] N. Blažić, and P. Gilkey, Conformally Osserman manifolds and conformally complex space forms, *Int. J. Geom. Methods Mod. Phys.* **1** (2004), 97–106. DOI: 10.1142/S021988780400006X

[34] N. Blažić, and P. Gilkey, Conformally Osserman manifolds and self-duality in Riemannian geometry, *Differential Geometry and its Applications*, 15–18, Matfyzpress, Prague, 2005.

[35] N. Blažić, P. Gilkey, S. Nikčević, and U. Simon, The spectral geometry of the Weyl conformal tensor, *Banach Center Publ.* **69** (2005), 195–203. DOI: 10.4064/bc69-0-15

[36] N. Blažić, P. Gilkey, S. Nikčević, and U. Simon, Algebraic theory of affine curvature tensors, *Arch. Math. (Brno)*, **42** (2006), Supplement, 147–168.

[37] N. Blažić, P. Gilkey, S. Nikčević, and I. Stavrov, Curvature structure of self-dual 4-manifolds, *Int. J. Geom. Methods Mod. Phys.* **7**, 1191-1204. DOI: 10.1142/S0219887808003259

[38] E. Boeckx, O. Kowalski, and L. Vanhecke, Riemannian manifolds of conullity two, *World Sci. Publ. Co.*, River Edge, NJ, 1996.

[39] N. Bokan, On the complete decomposition of curvature tensors of Riemannian manifolds with symmetric connection, *Rend. Circ. Mat. Palermo* **39** (1990), 331–380. DOI: 10.1007/BF02844767

[40] N. Bokan, M. Djorić, and U. Simon, Geometric structures as determined by the volume of generalized geodesic balls, *Results Math.* **43** (2003), 205–234.

[41] A. Bonome, P. Castro, and E. García-Río, Generalized Osserman four-dimensional manifolds, *Classical Quantum Gravity* **18** (2001), 4813–4822. DOI: 10.1088/0264-9381/18/22/307

[42] A. Bonome, R. Castro, E. García-Río, L. Hervella, and Y. Matsushita, Pseudo-Chern classes and opposite Chern classes of indefinite almost Hermitian manifolds, *Acta Math. Hungar.* **75** (1997), 299–316. DOI: 10.1023/A:1006545621874

[43] A. Bonome, R. Castro, E. García-Río, L. Hervella, and R. Vázquez-Lorenzo, Nonsymmetric Osserman indefinite Kähler manifolds, *Proc. Amer. Math. Soc.* **126** (1998), 2763–2769. DOI: 10.1090/S0002-9939-98-04659-0

[44] A. Bonome, R. Castro, E. García-Río, L. Hervella, and R. Vázquez-Lorenzo, Pseudo-Riemannian manifolds with simple Jacobi operators, *J. Math. Soc. Japan* **54** (2002), 847–875. DOI: 10.2969/jmsj/1191591994

[45] A. Borowiec, M. Francaviglia, and I. Volovich, Anti-Kählerian manifolds, *Differential Geom. Appl.* **12** (2000), 281–289. DOI: 10.1016/S0926-2245(00)00017-6

[46] C. P. Boyer, A note on hyper-Hermitian four-manifolds, *Proc. Amer. Math. Soc.* **102** (1988), 157–164. DOI: 10.2307/2046051

[47] C. P. Boyer, J. D. Finley III, and J. F. Plebański, Complex general relativity, ℌ and ℌℌ spaces – a survey of one approach, *Gen. Relativity Gravitation* **2** 241–281, Plenum, New York-London, 1980.

[48] M. Brozos-Vázquez, B. Fiedler, E. García-Río, P. Gilkey, S. Nikčević, G. Stanilov, Y. Tsankov, R. Vázquez-Lorenzo, and V. Videv, Stanilov-Tsankov-Videv Theory, *SIGMA Symmetry Integrability Geom. Methods Appl.* **3** (2007), 095, 13 pages. DOI: 10.3842/SIGMA.2007.095

[49] M. Brozos-Vázquez, E. García-Río, P. Gilkey, and R. Vázquez-Lorenzo, Compact Osserman metrics with neutral metric, preprint.

[50] M. Brozos-Vázquez, E. García-Río, P. Gilkey, and R. Vázquez-Lorenzo, Examples of signature (2, 2)-manifolds with commuting curvature operators, *J. Phys. A* **40** (2007), 13149–13159. DOI: 10.1088/1751-8113/40/43/021

[51] M. Brozos-Vázquez, E. García-Río, P. Gilkey, and R. Vázquez-Lorenzo, Completeness, Ricci blowup, the Osserman and the conformal Osserman condition for Walker signature (2, 2) manifolds, *Proceedings of XV International Workshop on Geometry and Physics*, Publ. de la RSME **10** (2007), 57–66.

[52] M. Brozos-Vázquez, E. García-Río, and R. Vázquez-Lorenzo, Conformally Osserman four-dimensional manifolds whose conformal Jacobi operators have complex eigenvalues, *Proc. R. Soc. A.* **462** (2006), 1425–1441. DOI: 10.1098/rspa.2005.1621

[53] M. Brozos-Vázquez, and P. Gilkey, Pseudo-Riemannian manifolds with commuting Jacobi operators, *Rend. Circ. Mat. Palermo* **55** (2006), 163–174. DOI: 10.1007/BF02874699

[54] M. Brozos-Vázquez, and P. Gilkey, The global geometry of Riemannian manifolds with commuting curvature operators, *J. Fixed Point Theory Appl.* **1** (2007), 87–96. DOI: 10.1007/s11784-006-0001-6

[55] M. Brozos-Vázquez, and P. Gilkey, Manifolds with commuting Jacobi operators, *J. Geom.* **86** (2007), 21–30. DOI: 10.1007/s00022-006-1898-z

[56] M. Brozos-Vázquez, P. Gilkey, H. Kang, S. Nikčević, and G. Weingart, Geometric realizations of curvature models by manifolds with constant scalar curvature, arxiv:0811.1651. To appear *Differential Geom. Appl.*

[57] M. Brozos-Vázquez, P. Gilkey, H. Kang, and S. Nikčević, Geometric Realizations of Hermitian curvature models, arxiv:0812.2743. To appear *J. Math. Soc. Japan*

[58] M. Brozos-Vázquez, P. Gilkey, and S. Nikčević, Jacobi–Tsankov manifolds which are not 2-step nilpotent, *Contemporary Geometry and Related Topics*, Belgrade (2005), 63–80.

[59] M. Brozos-Vázquez, P. Gilkey, S. Nikčević, and U. Simon, Projectively Osserman manifolds, *Publ. Math. Debrecen* **72** (2008), 359–370.

[60] M. Brozos-Vázquez, P. Gilkey, S. Nikčević, and R. Vázquez-Lorenzo, Geometric Realizations of para-Hermitian curvature models, arxiv:0902.1697. To appear *Results Math.*

[61] R. L. Bryant, Bochner-Kähler metrics, *J. Amer. Math. Soc.* **14** (2001), 623–715. DOI: 10.1090/S0894-0347-01-00366-6

[62] P. Bueken, Three-dimensional Lorentzian manifolds with constant principal Ricci curvatures $\rho_1 = \rho_2 \neq \rho_3$, *J. Math. Phys.* **38** (1997), 1000–1013. DOI: 10.1063/1.531880

[63] P. Bueken, On curvature homogeneous three-dimensional Lorentzian manifolds, *J. Geom. Phys.* **22** (1997), 349–362. DOI: 10.1016/S0393-0440(96)00037-X

[64] P. Bueken, and M. Djorić, Three-dimensional Lorentz metrics and curvature homogeneity of order one, *Ann. Global Anal. Geom.* **18** (2000), 85–103. DOI: 10.1023/A:1006612120550

[65] P. Bueken, and L. Vanhecke, Examples of curvature homogeneous Lorentz metrics, *Classical Quantum Gravity* **14** (1997), L93–L96. DOI: 10.1088/0264-9381/14/5/004

[66] M. Cahen, J. Leroy, M. Parker, F. Tricerri, and L. Vanhecke, Lorentz manifolds modeled on a Lorentz symmetric space, *J. Geom. Phys.* **7** (1990), 571–581. DOI: 10.1016/0393-0440(90)90007-P

[67] G. Calvaruso, Homogeneous structures on three-dimensional Lorentzian manifolds, *J. Geom. Phys.* **57** (2007), 1279–1291. DOI: 10.1016/j.geomphys.2006.10.005

[68] G. Calvaruso, Einstein-like metrics on three-dimensional homogeneous Lorentzian manifolds, *Geom. Dedicata* **127** (2007), 99–119. DOI: 10.1007/s10711-007-9163-7

[69] G. Calvaruso, Three-dimensional homogeneous Lorentzian metrics with prescribed Ricci tensor, *J. Math. Phys.* **48** (2007), no. 12, 123518, 17 pp. DOI: 10.1063/1.2825176

[70] G. Calvaruso, Pseudo-Riemannian 3-manifolds with prescribed distinct constant Ricci eigenvalues, *Differential Geom. Appl.* **26** (2008), 419–433. DOI: 10.1016/j.difgeo.2007.11.031

[71] G. Calvaruso, Addendum to Homogeneous structures on three-dimensional Lorentzian manifolds, *J. Geom. Phys.* **58** (2008), 291–292. DOI: 10.1016/j.geomphys.2007.10.006

[72] G. Calvaruso, Curvature homogeneous Lorentzian three-manifolds, to appear *Ann. Glob. Anal. Geom.*. DOI: 10.1007/s10455-008-9144-6

[73] G. Calvaruso, Semi-symmetric Lorentzian metrics and three-dimensional curvature homogeneity of order one, to appear *Abh. Math. Sem. Univ. Hamburg*.

[74] G. Calvaruso, Einstein-like Lorentz metrics and three-dimensional curvature homogeneity of order one, to appear *Canadian Math. Bull.*.

[75] E. Calviño-Louzao, E. García-Río, and R. Vázquez-Lorenzo, Four dimensional Osserman Ivanov–Petrova metrics of neutral signature, *Classical Quantum Gravity* **24** (2007), 2343–2355. DOI: 10.1088/0264-9381/24/9/012

[76] E. Calviño-Louzao, E. García-Río, and R. Vázquez-Lorenzo, Riemann extensions of torsion-free connections with degenerate Ricci tensor, to appear *Canad. J. Math.*.

[77] E. Calviño-Louzao, E. García-Río, P. Gilkey, and R. Vázquez-Lorenzo, The geometry of modified Riemannian extensions, *Proc. R. Soc. A.* (2009), doi:10.1098/rspa.2009.0046.

[78] M. do Carmo, *Riemannian geometry*, Birkhäuser Boston, Inc., Boston, MA, (1992).

[79] M. Chaichi, E. García-Río, and Y. Matsushita, Curvature properties of four-dimensional Walker metrics, *Classical Quantum Gravity* **22** (2005), 559–577. DOI: 10.1088/0264-9381/22/3/008

[80] M. Chaichi, E. García-Río, and M. E. Vázquez-Abal, Three-dimensional Lorentz manifolds admitting a parallel null vector field, *J. Phys. A* **38** (2005), 841–850. DOI: 10.1088/0305-4470/38/4/005

[81] M. C. Chaki, and B. Gupta, On conformally symmetric spaces, *Indian J. Math.* **5** (1963), 113–122.

[82] Y-K. E. Cheung, Y. Oz, and Z. Yin, Families of $N = 2$ strings, *J. High Energy Phys.* **11** (2003), 026, 55pp.

[83] Q. S. Chi, A curvature characterization of certain locally rank-one symmetric spaces, *J. Differential Geom.* **28** (1988), 187–202.

[84] Q. S. Chi, Curvature characterization and classification of rank-one symmetric spaces, *Pacific J. Math.* **150** (1991), 31–42.

[85] A. Chudecki, and M. Przanowski, From hyperheavenly spaces to Walker and Osserman spaces: I, *Classical Quantum Gravity* **25** (2008), 145010, 18pp. DOI: 10.1088/0264-9381/25/14/145010

[86] A. Chudecki, and M. Przanowski, From hyperheavenly spaces to Walker and Osserman spaces: II, *Classical Quantum Gravity* **25** (2008), 235019, 22pp. DOI: 10.1088/0264-9381/25/14/145010

[87] L. A. Cordero, and P. Parker, Left-invariant Lorentzian metrics on 3-dimensional Lie groups, *Rend. Mat.* **17** (1997), 129–155.

[88] L. Cordero, and P. Parker, Lattices and periodic geodesics in pseudoriemannian 2-step nilpotent Lie groups, *Int. J. Geom. Methods Mod. Phys.* **5** (2008), 79–99. DOI: 10.1142/S0219887808002667

[89] L. A. Cordero, and P. Parker, Pseudo Riemannian 2-step nilpotent Lie groups, arxiv.org/abs/math.DG/9905188.

[90] V. Cortés, C. Mayer, T. Mohaupt, and F. Saueressig, Special Geometry of Euclidean Supersymmetry. I.: Vector Multiplets, *J. High Energy Phys.* **23** (2004), 028, 73 pp.

[91] A. Cortés-Ayaso, J. C. Díaz-Ramos, and E. García-Río, Four-dimensional manifolds with degenerate self-dual Weyl curvature tensor, *Ann. Global Anal. Geom.* **34** (2008), 185–193. DOI: 10.1007/s10455-007-9101-9

[92] V. Cruceanu, P. Fortuny, and P. M. Gadea, A survey on paracomplex geometry, *Rocky Mountain J. Math.* **26** (1996), 83–115. DOI: 10.1216/rmjm/1181072105

[93] J. Davidov, J. C. Díaz-Ramos, E. García-Río, Y. Matsushita, O. Muskarov, and R. Vázquez-Lorenzo, Almost Kähler Walker 4-manifolds, *J. Geom. Phys.* **57** (2007), 1075–1088. DOI: 10.1016/j.geomphys.2006.09.003

[94] J. Davidov, J. C. Díaz-Ramos, E. García-Río, Y. Matsushita, O. Muskarov, and R. Vázquez-Lorenzo, Hermitian Walker 4-manifolds, *J. Geom. Phys.* **58** (2008) 307–323. DOI: 10.1016/j.geomphys.2007.11.006

[95] J. Davidov, and O. Muškarov, Self-dual Walker metrics with a two-step nilpotent Ricci operator, *J. Geom. Phys.* **57** (2006), 157–165. DOI: 10.1016/j.geomphys.2006.02.007

[96] V. de Smedt, Decomposition of the curvature tensor of hyper-Kaehler manifolds, *Lett. Math. Phys.* **30** (1994), 105–117. DOI: 10.1007/BF00939698

[97] A. Derdzinski, Self-dual Kähler manifolds and Einstein manifolds in dimension 4, *Compositio Math.* **49** (1983), 405–433.

[98] A. Derdzinski, Einstein metrics in dimension four, Handbook of Differential Geometry. Vol. I, (2000), 419–707, North-Holland, Amsterdam.

[99] A. Derdzinski, Curvature-homogeneous indefinite Einstein metrics in dimension four: the diagonalizable case, *Contemp. Math.* **337** (2003), 21–38.

[100] A. Derdzinski, Connections with skew-symmetric Ricci tensor on surfaces, *Results Math.* **52** (2008), 223–245. DOI: 10.1007/s00025-008-0307-3

[101] A. Derdzinski, Non-Walker self-dual neutral Einstein four-manifolds of Petrov Type III, *J. Geom. Anal.* **19** (2009), 301–357. DOI: 10.1007/s12220-008-9066-3

[102] A. Derdzinski, and W. Roter, On conformally symmetric manifolds with metrics of indices 0 and 1, *Tensor N. S.* **31** (1977), 255–259.

[103] A. Derdzinski, and W. Roter, Walker's theorem without coordinates, *J. Math. Phys.* **47** (2006) 062504, 8 pp. DOI: 10.1063/1.2209167

[104] A. Derdzinski, and W. Roter, The local structure of conformally symmetric manifolds, to appear *Bull. Belg. Math. Soc. Simon Stevin.*

[105] J. C. Díaz-Ramos, and E. García-Río, A note on the structure of algebraic curvature tensors, *Linear Alg. Appl.* **382** (2004), 271–277. DOI: 10.1016/j.laa.2003.12.044

[106] J. C. Díaz-Ramos, E. García-Río, and R. Vázquez-Lorenzo, Four dimensional Osserman metrics with nondiagonalizable Jacobi operators, *J. Geom. Anal.* **16** (2006), 39–52. DOI: 10.1007/BF02930986

[107] J. C. Díaz-Ramos, E. García-Río, and R. Vázquez-Lorenzo, New examples of Osserman metrics with nondiagonalizable Jacobi operators, *Differential Geom. Appl.* **24** (2006), 433–442. DOI: 10.1016/j.difgeo.2006.02.006

[108] J. C. Díaz-Ramos, E. García-Río, and R. Vázquez-Lorenzo, Osserman metrics on Walker 4-manifolds equipped with a para-Hermitian structure, *Mat. Contemp.* **30** (2006), 91–108.

[109] S. Dragomir, and K. L. Duggal, Indefinite locally conformal Kaehler manifolds, *Differential Geom. Appl.* **25** (2007), 8–22. DOI: 10.1016/j.difgeo.2006.11.002

[110] S. Dragomir, and L. Ornea, Locally conformally Kaehler geometry, *Progress in Mathematics* **155**, Birkhäuser, Boston, 1998.

[111] M. Dunajski, Anti-self-dual four-manifolds with a parallel real spinor, *Proc. R. Soc. A.* **458** (2002), 1205–1222. DOI: 10.1098/rspa.2001.0918

[112] M. Dunajski, and S. West, Anti-Self-Dual Conformal Structures in Neutral Signature, *Recent developments in pseudo-Riemannian geometry*, 113–148, ESI Lect. Math. Phys., Eur. Math. Soc., Zürich, 2008.

[113] C. Dunn, and P. Gilkey, Curvature homogeneous pseudo-Riemannian manifolds which are not locally homogeneous, *Complex, contact and symmetric manifolds*, 145–152, Progr. Math., **234**, Birkhäuser Boston, Boston, MA, 2005

[114] C. Dunn, P. Gilkey, and S. Nikčević, Curvature homogeneous signature (2, 2) manifolds, *Differential geometry and its applications*, 29–44, Matfyzpress, Prague, 2005.

[115] D. Ferus, H. Karcher, and H. Münzner, Cliffordalgebren und neue isoparametrische Hyperflächen, *Math. Z.* **177** (1981), 479–502. DOI: 10.1007/BF01219082

[116] A. Fialkow, Hypersurfaces of a space of constant curvature, *Ann. of Math. (2)* **39** (1938), 762–785. DOI: 10.2307/1968462

[117] B. Fiedler, Determination of the structure of algebraic curvature tensors by means of Young symmetrizers, Seminaire *Sém. Lothar. Combin.* **48** (2002), Art. B48d, 20 pp

[118] B. Fiedler, and P. Gilkey, Nilpotent Szabó, Osserman and Ivanov–Petrova pseudo-Riemannian manifolds, *Contemp. Math.* **337** (2003), 53–63.

[119] J. D. Finley III, and J. F. Plebański, The intrinsic spinorial structure of hyperheavens, *J. Math. Phys.* **17** (1976), 2207–2214. DOI: 10.1063/1.522867

[120] E. J. Flaherty, *Hermitian and Kählerian geometry in relativity*, Lect. Notes in Phys. **46**, Springer-Verlag, Berlin-New York, 1976.

[121] P. M. D. Furness, and S. A. Robertson, Parallel framings and foliations on pseudoriemannian manifolds, *J. Differential Geom.* **9** (1974), 409–422.

[122] P. M. Gadea, and J. A. Oubiña, Homogeneous pseudo-Riemannian structures and homogeneous almost para-Hermitian structures, *Houston J. Math.* **18** (1992), 449–465.

[123] P. M. Gadea, and J. A. Oubiña, Reductive homogeneous pseudo-Riemannian manifolds, *Monatsh. Math.* **124** (1997), 17–34. DOI: 10.1007/BF01320735

[124] E. García-Río, P. Gilkey, M. E. Vázquez-Abal, and R. Vázquez-Lorenzo, Four dimensional Osserman metrics of neutral signature, preprint (arxiv.org/abs/0804.0436v1).

[125] E. García-Río, A. Haji-Badali, M. E. Vázquez-Abal, and R. Vázquez-Lorenzo, Lorentzian 3 manifolds with commuting curvature operators, *Int. J. Geom. Methods Mod. Phys.* **5** (2008), 557–572. DOI: 10.1142/S0219887808002941

[126] E. García-Río, A. Haji-Badali, M. E. Vázquez-Abal, and R. Vázquez-Lorenzo, On the local geometry of the three-dimensional Walker metrics, Advances in Lorentzian Geometry (M. Plaue, M. Scherfner, eds.), 77-87, Shaker Verlag, Berlin 2008.

[127] E. García-Río, A. Haji-Badali, and R. Vázquez-Lorenzo, Lorentzian three-manifolds with special curvature operators, *Classical Quantum Gravity* **25** (2008) 015003, 13pp. DOI: 10.1088/0264-9381/25/1/015003

[128] E. García-Río, D. Kupeli, and M. E. Vázquez-Abal, On a problem of Osserman in Lorentzian geometry, *Differential Geom. Appl.* **7** (1997), 85–100. DOI: 10.1016/S0926-2245(96)00037-X

[129] E. García-Río, D. Kupeli, M. E. Vázquez-Abal, and R. Vázquez-Lorenzo, Affine Osserman connections and their Riemann extensions, *Differential Geom. Appl.* **11** (1999), 145–153. DOI: 10.1016/S0926-2245(99)00029-7

[130] E. García-Río, D. Kupeli, and R. Vázquez-Lorenzo, Osserman manifolds in semi-Riemannian geometry, Lect. Notes Math. **1777**, *Springer-Verlag*, Berlin, Heidelberg, New York, 2002.

[131] E. García-Río, and Y. Matsushita, Isotropic Kaehler structures on Engel 4-manifolds, *J. Geom. Phys.* **33** (2000), 288–294. DOI: 10.1016/S0393-0440(99)00055-8

[132] E. García–Río, Z. Rakić, and M. E. Vázquez-Abal, Four-dimensional indefinite Kähler Osserman manifolds, *J. Math. Phys.* **46** (2005), 11pp. DOI: 10.1063/1.1938727

[133] E. García-Río, M. E. Vázquez-Abal, and R. Vázquez-Lorenzo, Nonsymmetric Osserman pseudo-Riemannian manifolds, *Proc. Amer. Math. Soc.* **126** (1998), 2771–2778. DOI: 10.1090/S0002-9939-98-04666-8

[134] E. García-Río, and R. Vázquez-Lorenzo, Four-dimensional Osserman symmetric spaces, *Geom. Dedicata* **88** (2001), 147–151. DOI: 10.1023/A:1013101719550

[135] P. Giblin, Obituary: Arthur Geoffrey Walker 1909–2001, *Bull. London Math. Soc.* **36** (2004), 271–280. DOI: 10.1112/S0024609303002765

[136] P. Gilkey, Manifolds whose curvature operator has constant eigenvalues at the basepoint, *J. Geom. Anal.* **4** (1994), 155–158. DOI: 10.1007/BF02921544

[137] P. Gilkey, Geometric properties of natural operators defined by the Riemannian curvature tensor, *World Scientific Publishing Co., Inc.*, River Edge, NJ, 2001.

[138] P. Gilkey, Algebraic curvature tensors which are p-Osserman, *Differential Geom. Appl.* **14** (2001), 297–311. DOI: 10.1016/S0926-2245(01)00040-7

[139] P. Gilkey, Bundles over projective spaces and algebraic curvature tensors, *J. Geom* **71** (2001), 54–67. DOI: 10.1007/s00022-001-8552-6

[140] P. Gilkey, The geometry of curvature homogeneous pseudo-Riemannian manifolds. *ICP Advanced Texts in Mathematics*, **2**. Imperial College Press, London, 2007.

[141] P. Gilkey, and R. Ivanova, The Jordan normal form of Osserman algebraic curvature tensors, *Results Math.* **40** (2001), 192–204.

[142] P. Gilkey, and R. Ivanova, Spacelike Jordan-Osserman algebraic curvature tensors in the higher signature setting, *Differential Geometry, Valencia, 2001*, World Sci. Publ., River Edge, NJ, 2002, 179–186.

[143] P. Gilkey, R. Ivanova, and I. Stavrov, Jordan Szabó algebraic covariant derivative curvature tensors, *Contemp. Math.* **337** (2003), 65–75.

[144] P. Gilkey, R. Ivanova, and T. Zhang, Szabó Osserman IP pseudo Riemannian manifolds *Publ. Math. Debrecen* **62** (2003), 387–401.

[145] P. Gilkey, J. V. Leahy, and H. Sadofsky, Riemannian manifolds whose skew-symmetric curvature operator has constant eigenvalues, *Indiana Univ. Math. J.* **48** (1999), 615–634. DOI: 10.1512/iumj.1999.48.1699

[146] P. Gilkey, and S. Nikčević, Complete curvature homogeneous pseudo-Riemannian manifolds, *Classical Quantum Gravity*, **21** (2004), 3755–3770. DOI: 10.1088/0264-9381/21/15/009

[147] P. Gilkey, and S. Nikčević, Curvature homogeneous spacelike Jordan Osserman pseudoRiemannian manifolds, *Classical Quantum Gravity* **21** (2004), 497–507. DOI: 10.1088/0264-9381/21/2/013

[148] P. Gilkey, and S. Nikčević, Generalized plane wave manifolds, *Kragujevac J. Math.* **28** (2005), 113–138.

[149] P. Gilkey, and S. Nikčević, Complete k-curvature homogeneous pseudo-Riemannian manifolds, *Ann. Global Anal. Geom.* **27** (2005), 87–100. DOI: 10.1007/s10455-005-5217-y

[150] P. Gilkey, and S. Nikčević, Pseudo-Riemannian Jacobi-Videv manifolds, *Int. J. Geom. Methods Mod. Phys.* **4** (2007), 727–738. DOI: 10.1142/S0219887807002272

[151] P. Gilkey, and S. Nikčević, Geometrical representations of equiaffine curvature operators, *Results Math.* **52** (2008), 281–287. DOI: 10.1007/s00025-008-0310-8

[152] P. Gilkey, and S. Nikčević, The classification of simple Jacobi–Ricci commuting algebraic curvature tensors, to appear *Note di Matematica*.

[153] P. Gilkey, S. Nikčević, and V. Videv, Manifolds which are Ivanov-Petrova or k-Stanilov, *J. Geom.* **80** (2004), 82–94. DOI: 10.1007/s00022-003-1750-7

[154] P. Gilkey, S. Nikčević, and D. Westerman, Geometric realizations of generalized algebraic curvature operators, *J. Math. Phys.* **50** (2009), 013515. DOI: 10.1063/1.3049619

[155] P. Gilkey, G. Stanilov, and V. Videv, Pseudo-Riemannian manifolds whose generalized Jacobi operator has constant characteristic polynomial, *J. Geom.* **62** (1988), 144–153. DOI: 10.1007/BF01237606

[156] P. Gilkey, and I. Stavrov, Curvature tensors whose Jacobi or Szabó operator is nilpotent on null vectors, *Bull. London Math. Soc.* **34** (2002), 650–658. DOI: 10.1112/S0024609302001339

[157] P. Gilkey, A. Swann, and L. Vanhecke, Isoparametric geodesic spheres and a conjecture of Osserman concerning the Jacobi operator, *Quart. J. Math. Oxford Ser. (2)* **46** (1995), 299–320. DOI: 10.1093/qmath/46.3.299

[158] P. Gilkey, and V. Videv, Jacobi–Jacobi commuting models and manifolds, *J. Geom.* **92** (2009), 60–68. DOI: 10.1007/s00022-008-2061-9

[159] P. Gilkey, and T. Zhang, Algebraic curvature tensors for indefinite metrics whose skew-symmetric curvature operator has constant Jordan normal form, *Houston J. Math.* **28** (2002), 311–328.

[160] S. Goldberg, Integrability of almost Kaehler manifolds, *Proc. Amer. Math. Soc.* **21** (1969), 96–100. DOI: 10.2307/2036867

[161] A. Gray, Curvature identities for Hermitian and almost Hermitian manifolds, *Tôhoku Math. J.* **28** (1976), 601–612. DOI: 10.2748/tmj/1178240746

[162] A. Gray, Einstein-like manifolds which are not Einstein, *Geom. Dedicata* **7** (1978), 259–280. DOI: 10.1007/BF00151525

[163] A. Gray, and L. M. Hervella, The sixteen classes of almost Hermitian manifolds and their linear invariants, *Ann. Mat. Pure Appl.* **123** (1980), 35–58. DOI: 10.1007/BF01796539

[164] M. Gromov, *Partial differential relations*, Ergeb. Math. Grenzgeb 3. Folge, Band 9, Springer-Verlag (1986).

[165] G. S. Hall, Covariantly constant tensors and holonomy structure in general relativity, *J. Math. Phys.* **32** (1991), 181–187. DOI: 10.1063/1.529114

[166] G. S. Hall, and J. da Costa, Affine collineations in space-time, *J. Math. Phys.* **29** (1988), 2465–2472. DOI: 10.1063/1.528083

[167] L. Hernández-Lamoneda, Curvature vs. almost Hermitian structures, *Geom. Dedicata* **79** (2000), 205–218. DOI: 10.1023/A:1005232107795

[168] N. Hitchin, Hypersymplectic quotients, *Acta Acad. Sci. Tauriensis* **124** supl., (1990), 169–180.

[169] K. Honda, Conformally flat semi-Riemannian manifolds with commuting curvature and Ricci operators, *Tokyo J. Math.* **26** (2003), 241–260.

[170] K. Honda, and K. Tsukada, Three-dimensional conformally flat homogeneous Lorentzian manifolds, *J. Math. Phys.* **40** (2007), 831–851. DOI: 10.1088/1751-8113/40/4/017

[171] S. Ivanov, and I. Petrova, Riemannian manifolds in which certain curvature operator has constant eigenvalues along each circle, *Ann. Global Anal. Geom.* **15** (1997), 157–171. DOI: 10.1023/A:1006548328030

[172] S. Ivanov, and I. Petrova, Riemannian manifold in which the skew-symmetric curvature operator has pointwise constant eigenvalues, *Geom. Dedicata* **70** (1998), 269–282. DOI: 10.1023/A:1005014507809

[173] S. Ivanov, and S. Zamkovoy, Parahermitian and paraquaternionic manifolds, *Differential Geom. Appl.* **23** (2005), 205–234. DOI: 10.1016/j.difgeo.2005.06.002

[174] M. Ivanova, and V. Videv, Four-dimensional Riemannian manifolds with commuting Stanilov curvature operators, *Mathematics and education in mathematics*, Union of Bulgarian Mathematicians, Sofia, (2004), 184–188.

[175] M. Ivanova, V. Videv, and Z. Zhelev, Four-dimensional Riemannian manifolds with commuting higher order Jacobi operators, *Plovdiv University "Paisii Hilendarski", Bulgaria, Scientific Works* **35** Book 3 (2007), 167–180.

[176] A. Jevicki, M. Mihailescu, and J. P. Nunes, Large N field theory of $N =$ 2 strings and self-dual gravity, *Chaos Solitons Fractals* **10** (1999), 385–397. DOI: 10.1016/S0960-0779(98)00213-6

[177] H. Kamada, Self-duality of neutral metrics on four-manifolds, *Monogr. Geom. Topology* **25**, 79–98, Int. Press, Cambridge, MA, 1998.

[178] H. Kamada, Neutral hyperkähler structures on primary Kodaira surfaces, *Tsukuba J. Math.* **23** (1999), 321–332.

[179] J. Kerimo, AdS pp-waves, *J. High Energy Phys.* (2005), 025, 18 pp. DOI: 10.1088/1126-6708/2005/09/025

[180] K.-D. Kirchberg, Integrability conditions for almost Hermitian and almost Kaehler 4-manifolds, arxiv:math.DG/0605611.

[181] J. Klusoň, R. I. Nayak, and K. L. Panigrahi, D-brane dynamics in a plane wave background, *Phys. Rev. D* **73** (2006), no. 6, 066007, 10 pp. DOI: 10.1103/PhysRevD.73.066007

[182] S. Kobayashi, and K. Nomizu, Foundations of Differential Geometry I, *Interscience Publ.*, New York, 1963.

[183] A. Koutras, and C. McIntosh, A metric with no symmetries or invariants, *Classical Quantum Gravity* **13** (1996), L47–L49. DOI: 10.1088/0264-9381/13/5/002

[184] O. Kowalski, B. Opozda, and Z. Vlášek, A classification of locally homogeneous affine connections with skew-symmetric Ricci tensor on 2-dimensional manifolds, *Monatsh. Math.* **130** (2000), 109–125. DOI: 10.1007/s006050070041

[185] O. Kowalski, B. Opozda, and Z. Vlášek, A classification of locally homogeneous connections on 2-dimensional manifolds via group-theoretical approach, *Cent. Eur. J. Math.* **2** (2004), 87–102. DOI: 10.2478/BF02475953

[186] J. Lafontaine, Conformal geometry from the Riemannian viewpoint, *Conformal geometry (Bonn, 1985/1986)* Aspects Math. **E12** (1988) Vieweg Braunschweig, 65–92.

[187] P. R. Law, Neutral Einstein metrics in four dimensions, *J. Math. Phys.* **32** (1991), 3039–3042. DOI: 10.1063/1.529048

[188] P. R. Law, and Y. Matsushita, Hitchin-Thorpe-type inequalities for pseudo-Riemannian 4-manifolds of metric signature $(+ + - -)$, *Geom. Dedicata* **87** (2001), 65–89. DOI: 10.1023/A:1012002211862

[189] P. R. Law, and Y. Matsushita, A Spinor Approach to Walker Geometry, *Comm. Math. Phys.* **282** (2008), 577–623. DOI: 10.1007/s00220-008-0561-y

[190] Th. Leistner, Screen bundles of Lorentzian manifolds and some generalizations of pp-waves, *J. Geom. Phys.* **56** (2006), 2117–2134. DOI: 10.1016/j.geomphys.2005.11.010

[191] A. M. Li, U. Simon, and G. Zhao, Global affine differential geometry of hypersurfaces, *de Gruyter Expositions in Mathematics* **11** (1993).

[192] M. A. Magid, Shape operators of Einstein hypersurfaces in indefinite space forms, *Proc. Amer. Math. Soc.* **84** (1982), 237–242. DOI: 10.2307/2043672

[193] M. A. Magid, Indefinite Einstein hypersurfaces with imaginary principal curvatures, *Houston J. Math.* **10** (1984), 57–61.

[194] M. A. Magid, Indefinite Einstein hypersurfaces with nilpotent shape operators, *Hokkaido Math. J.* **13** (1984), 241–250.

[195] N. Marcus, The $N = 2$ open string, *Nuclear Phys. B* **387** (1992), 263–279. DOI: 10.1016/0550-3213(92)90161-4

[196] A. Marden, Outer circles. An introduction to hyperbolic 3-manifolds, *Cambridge University Press*, Cambridge, 2007.

[197] Y. Matsushita, Four-dimensional Walker metrics and symplectic structures, *J. Geom. Phys.* **52** (2004), 89–99; see also Erratum *J. Geom. Phys.* **57** (2007), 729. DOI: 10.1016/j.geomphys.2004.02.009

[198] Y. Matsushita, Walker 4-manifolds with proper almost complex structures, *J. Geom. Phys.* **55** (2005), 385–398. DOI: 10.1016/j.geomphys.2004.12.014

[199] Y. Matsushita, S. Haze, and P. R. Law, Almost Kaehler-Einstein structures on 8-dimensional Walker manifolds, *Monatsh. Math.* **150** (2007), 41–48. DOI: 10.1007/s00605-006-0403-y

[200] J. Michelson, and X. Wu, Dynamics of antimembranes in the maximally supersymmetric eleven-dimensional pp wave, *J. High Energy Phys.* (2006), 028, 37 pp. (electronic). DOI: 10.1088/1126-6708/2006/01/028

[201] J. Milnor, Curvatures of left invariant metrics on Lie groups, *Adv. Math.* **21** (1976), 293–329. DOI: 10.1016/S0001-8708(76)80002-3

[202] R. Milson, and N. Pelavas, The curvature homogeneity bound for Lorentzian four manifolds, arxiv:0711.3851v2. DOI: 10.1142/S0219887809003424

[203] A. Montesinos Amilibia, Degenerate homogeneous structures of type $\mathcal{S}_1$ on pseudo-Riemannian manifolds, *Rocky Mountain J. Math.* **31** (2001), 561–579. DOI: 10.1216/rmjm/1020171575

[204] S. Montiel, and A. Romero, Complex Einstein hypersurfaces of indefinite complex space forms, *Math. Proc. Cambridge Philos. Soc.* **94** (1983), 495–508. DOI: 10.1017/S0305004100000888

[205] A. Newlander, and L. Nirenberg, Complex analytic coordinates in almost complex manifolds, *Ann. of Math.* **65** (1957), 391–404. DOI: 10.2307/1970051

[206] Y. Nikolayevsky, Osserman manifolds and Clifford structures, *Houston J. Math.* **29** (2003), 59–75.

[207] Y. Nikolayevsky, Riemannian manifolds whose curvature operator $R(X, Y)$ has constant eigenvalues, *Bull. Austral. Math. Soc.* **70** (2004), 301–319. DOI: 10.1017/S0004972700034523

[208] Y. Nikolayevsky, Osserman manifolds of dimension 8, *Manuscripta Math.* **115** (2004), 31–53. DOI: 10.1007/s00229-004-0480-y

[209] Y. Nikolayevsky, Osserman conjecture in dimension $\neq 8, 16$, *Math. Ann.* **331** (2005), 505–522.

[210] Y. Nikolayevsky, Conformally Osserman manifolds, arxiv:0810.5621.

[211] T. Oguro, and K. Sekigawa, Four-dimensional almost Kaehler Einstein and $\star$-Einstein manifolds, *Geom. Dedicata* **69** (1998), 91–112. DOI: 10.1023/A:1005005526324

[212] Z. Olszak, A note on almost-Kaehler manifolds, *Bull. Acad. Polon. Sci. Sér. Sci. Math. Astronom. Phys.* **26** (1978), 139–141.

[213] Z. Olszak, On conformally recurrent manifolds II. Riemann extensions, *Tensor N. S.* **49** (1990), 24–31.

[214] Z. Olszak, On conformally recurrent manifolds, I: Special distributions, *Zesz. Nauk. Politech. Sl., Mat.-Fiz.* **68** (1993), 213–225.

[215] B. O'Neill, Semi–Riemannian geometry, with applications to relativity, *Academic Press*, New York, 1983.

[216] H. Ooguri, and C. Vafa, Geometry of $N = 2$ strings, *Nuclear Phys.* **B 361** (1991), 469–518. DOI: 10.1016/0550-3213(91)90270-8

[217] B. Opozda, Affine versions of Singer's theorem on locally homogeneous spaces, *Ann. Global Anal. Geom.* **15** (1997), 187–199. DOI: 10.1023/A:1006585424144

[218] R. Osserman, Curvature in the eighties, *Amer. Math. Monthly* **97** (1990), 731–756. DOI: 10.2307/2324577

[219] J. A. Oubiña, New classes of almost contact metric structures, *Publ. Math. Debrecen* **32** (1985), 187–193.

[220] G. Ovando, Invariant pseudo Kaehler metrics in dimension four, *J. Lie Theory* **16** (2006), 371–391.

[221] E. M. Patterson, and A. G. Walker, Riemann extensions, *Quart. J. Math., Oxford Ser. (2)* **3** (1952), 19–28.

[222] H. Pedersen, and P. Tod, The Ledger curvature conditions and D'Atri geometry, *Differential Geom. Appl.* **11** (1999), 155–162. DOI: 10.1016/S0926-2245(99)00026-1

[223] J. Petean, Indefinite Kaehler-Einstein Metrics on compact complex surfaces, *Comm. Math. Phys.* **189** (1997), 227–235. DOI: 10.1007/s002200050197

[224] F. Podesta, and A. Spiro, Introduzione ai Gruppi di Trasformazioni, *Volume of the Preprint Series of the Mathematics Department "V. Volterra" of the University of Ancona*, Via delle Brecce Bianche, Ancona, ITALY (1996).

[225] V. Pravda, A. Pravdová, A. Coley, and R. Milson, All spacetimes with vanishing curvature invariants, *Classical Quantum Gravity* **19** (2002), 6213–6236. DOI: 10.1088/0264-9381/19/23/318

[226] F. Prüfer, F. Tricerri, and L. Vanhecke, Curvature invariants, differential operators and local homogeneity, *Trans. Amer. Math. Soc.* **348** (1996), 4643–4652. DOI: 10.1090/S0002-9947-96-01686-8

[227] S. Rahmani, Métriques de Lorentz sur les groupes de Lie unimodulaires de dimension trois, *J. Geom. Phys.* **9** (1992), 295–302. DOI: 10.1016/0393-0440(92)90033-W

[228] B. L. Reinhart, The second fundamental form of a plane field, *J. Differential Geom.* **12** (1977), 619–627.

[229] G. de Rham, Sur la réductibilité dùn espace de riemann, *Comment. Math. Helv.* **26** (1952), 328–344. DOI: 10.1007/BF02564308

[230] V. Sahni, and Y. Shtanov, New vistas in braneworld cosmology, *Internat. J. Modern Phys. D* **11** (2002), 1515–1521. DOI: 10.1142/S0218271802002827

[231] P. A. Schirokow, and A. P. Schirokow, Affine Differentialgeometrie, *Teubner Leipzig* (1962).

[232] K. Sekigawa, On some 3-dimensional curvature homogeneous spaces, *Tensor N.S.* **31** (1977), 87–97.

[233] K. Sekigawa, On some 4-dimensional compact Einstein almost Kaehler manifolds, *Math. Ann.* **271** (1985), 333–337. DOI: 10.1007/BF01456071

[234] K. Sekigawa, On some compact Einstein almost Kaehler manifolds, *J. Math. Soc. Japan* **39** (1987), 677–684. DOI: 10.2969/jmsj/03940677

[235] K. Sekigawa, H. Suga, and L. Vanhecke, Four-dimensional curvature homogeneous spaces, *Comment. Math. Univ. Carolin.* **33** (1992), 261–268.

[236] K. Sekigawa, H. Suga, and L. Vanhecke, Curvature homogeneity for four-dimensional manifolds, *J. Korean Math. Soc.* **32** (1995), 93–101.

[237] K. Sekigawa, and H. Takagi, On conformally flat spaces satisfying a certain condition on the Ricci tensor, *Tôhoku Math. J.* **23** (1971), 1–11. DOI: 10.2748/tmj/1178242681

[238] K. Sekigawa, and A. Yamada, Compact indefinite almost Kähler Einstein manifolds, *Geom. Dedicata* **132** (2008), 65–79. DOI: 10.1007/s10711-007-9166-4

[239] U. Simon, A. Schwenk-Schellschmidt, and H. Viesel, Introduction to the affine differential geometry of hypersurfaces, *Science University of Tokyo* 1991.

[240] I. M. Singer, Infinitesimally homogeneous spaces, *Commun. Pure Appl. Math.* **13** (1960), 685–697. DOI: 10.1002/cpa.3160130408

[241] I. M. Singer, and J. A. Thorpe, The curvature of 4-dimensional Einstein spaces, *1969 Global Analysis (Papers in Honor of K. Kodaira)*, Univ. Tokyo Press, Tokyo, 355–365.

[242] G. Stanilov, Higher order skew-symmetric and symmetric curvature operators, *C. R. Acad. Bulg. Sci.* **57** (2004), 9–12.

[243] G. Stanilov, Curvature operators based on the skew-symmetric curvature operator and their place in the Differential Geometry, preprint.

[244] G. Stanilov, and V. Videv, On a generalization of the Jacobi operator in the Riemannian geometry, *Annuaire Univ. Sofia Fac. Math. Inform.* **86** (1992), 27–34.

[245] G. Stanilov and V. Videv, On a generalization of the Jacobi operator in the Riemannian geometry, *God. Sofij. Univ., Fak. Mat. Inform.* **86** (1994) 27–34.

[246] G. Stanilov and V. Videv, On the commuting of curvature operators, *Mathematics and Education in Mathematics (Proc. of the 33rd Spring Conference of the Union of Bulgarian Mathematicians Borovtes, April 1-4, 2004)*, Sofia (2004), 176–179.

[247] R. Strichartz, Linear algebra of curvature tensors and their covariant derivatives, *Canad. J. Math.* **40** (1988), 1105–1143.

[248] Z. I. Szabó, A short topological proof for the symmetry of 2 point homogeneous spaces, *Invent. Math.* **106** (1991), 61–64. DOI: 10.1007/BF01243903

[249] H. Takagi, On curvature homogeneity of Riemannian manifolds, *Tôhoku Math. J.* **26** (1974), 581–585. DOI: 10.2748/tmj/1178241081

[250] T. Y. Thomas, The decomposition of Riemann spaces in the large, *Monatsh. Math. Phys.* **47** (1939), 388–418. DOI: 10.1007/BF01695510

[251] F. Tricerri, and L. Vanhecke, Curvature tensors on almost Hermitian manifolds, *Trans. Amer. Math. Soc.* **267** (1981), 365–397. DOI: 10.2307/1998660

[252] F. Tricerri, and L. Vanhecke, Homogeneous structures on Riemannian manifolds, *Cambridge University Press*, 1983.

[253] Y. Tsankov, A characterization of n-dimensional hypersurface in $\mathbb{R}^{n+1}$ with commuting curvature operators, *Banach Center Publ.* **69** (2005), 205–209. DOI: 10.4064/bc69-0-16

[254] A. G. Walker, On parallel fields of partially null vector spaces, *Quart. J. Math. Oxford* **20** (1949), 135–145. DOI: 10.1093/qmath/os-20.1.135

[255] A. G. Walker, Canonical form for a Riemannian space with a parallel field of null planes, *Quart. J. Math. Oxford* (2) **1** (1950), 69–79. DOI: 10.1093/qmath/1.1.69

[256] A. G. Walker, Sur la fibration des variétés riemanniennes, *C. R. Acad. Sci. Paris* **232** (1951), 1465–1467.

[257] A. G. Walker, The fibering of Riemannian manifolds, *Proc. London Math. Soc. (3)* **3** (1953), 1–19. DOI: 10.1112/plms/s3-3.1.1

[258] A. G. Walker, Connexions for parallel distributions in the large, *Quart. J. Math. Oxford (2)* **6** (1955), 301–308. DOI: 10.1093/qmath/6.1.301

[259] A. G. Walker, Connexions for parallel distributions in the large. II, *Quart. J. Math. Oxford (2)* **9** (1958), 221–231. DOI: 10.1093/qmath/9.1.221

[260] Y. C. Wong, Two dimensional linear connexions with zero torsion and recurrent curvature, *Monatsh. Math.* **68** (1964), 175–184. DOI: 10.1007/BF01307120

[261] H. Wu, On the de Rham decomposition theorem, *Illinois J. Math.* **8** (1964), 291–311.

[262] K. Yamato, Algebraic Riemann manifolds, *Nagoya Math. J.* **115** (1989), 87–104.

[263] K. Yano, and S. Ishihara, Tangent and cotangent bundles, *Pure and Applied Mathematics* **16** Marcel Dekker, New York, 1973.

[264] T. Zhang, Manifolds with indefinite metrics whose skew-symmetric curvature operator has constant eigenvalues Ph. D. Thesis University of Oregon, 2000.

Glossary

In order to help the reader to understand the basic content of the book, we summarize in these pages some notational conventions employed herein.

Elements in a model.

V	vector space.
$x, y, z,\ldots$	arbitrary vectors of V.
$\langle\cdot,\cdot\rangle$	inner product of arbitrary signature.
A	algebraic curvature tensor.
$\mathfrak{M}$	algebraic model composed by the triple $(V, \langle\cdot,\cdot\rangle, A)$.
$\mathfrak{C}$	complex model composed by the quadruple $(V, \langle\cdot,\cdot\rangle, J, A)$.
$\Xi(V)$	space of algebraic curvature tensors.

Elements in a manifold.

M	differentiable manifold of dimension m.
$\mathcal{O}$	open neighborhood.
g_D	Riemannian extension.
g	pseudo-Riemannian metric.
$\mathcal{M}$	pseudo-Riemannian manifold formed by the pair (M, g).
$(x_1,\ldots,x_m)$	local coordinates.
$C^\infty(M)$	space of smooth functions on M.
$T_P M$	tangent space of M at P.
$C^\infty(TM)$	space of smooth vector fields on M.
$\{\partial_{x_1},\ldots,\partial_{x_m}\}$	coordinate vector fields.
$\{dx_1,\ldots,dx_m\}$	dual of the coordinate vector fields.
$X, Y, Z,\ldots$	arbitrary vector fields of M.
D	arbitrary connection.
∇	Levi-Civita connection.

Curvature and associated elements.

R	$(0, 4)$ curvature tensor.
$\mathcal{R}$	curvature operator.
$R_{ijk}{}^{\ell}$	curvature components in a given basis.
ρ	Ricci tensor and Ricci operator.
τ	scalar curvature.
W^+, W^-	self-dual and anti-self-dual Weyl curvature tensors.
L	second fundamental form.

Complex and para-complex manifolds.

J	almost Hermitian and almost para-Hermitian structure.
$\{J_1, J_2, J_3\}$	hyper-Hermitian and hyper-para-Hermitian structure.
$\rho^{\star}$	$\star$-Ricci tensor.
$\tau^{\star}$	$\star$-scalar curvature.
$\Omega(\cdot, \cdot)$	Kaehler form.

Walker manifolds.

$\mathcal{D}$	null distribution.
$g_{D,\phi}$	twisted Riemannian extension.
$g_{D,\phi,T,S}$	modified Riemannian extension.
$\mathcal{M}_{a,b,c}$	Walker manifold of signature $(2, 2)$ composed of the pair $(\mathcal{O}, g_{a,b,c})$.
$\mathcal{C}_{a,b,c}$	almost Hermitian manifold of signature $(2, 2)$ composed of the triple $(\mathcal{O}, g_{a,b,c}, J)$.

Biography

MIGUEL BROZOS-VÁZQUEZ

Miguel Brozos-Vázquez got his Ph.D. in 2007 from the University of Santiago de Compostela under the direction of Eduardo García-Río and Ramón Vázquez-Lorenzo. In 2005 and 2006, respectively, he visited the University of Oregon and the Max-Planck-Institut für Mathematik in den Naturwissenschaften. After a year working as a teacher in Secondary School, he became Assistant Professor at the Universidade da Coruña in 2008. His research focusses mainly on Pseudo-Riemannian Geometry.

EDUARDO GARCÍA-RÍO

Eduardo García-Río received his Ph.D. from the University of Santiago de Compostela in 1992 under the direction of A. Bonome and L. Hervella. He is an Associate Professor of Geometry at the University of Santiago de Compostela since 2000. His research focuses on Differential Geometry and Global Analysis. He has organized several courses and conferences on Riemannian and Lorentzian geometry and is a foundational member of the Spanish Institute of Mathematics.

PETER GILKEY

Peter Gilkey received his Ph.D. from Harvard in 1972 under the direction of Louis Nirenberg. After holding positions at Berkeley, at Princeton, and at U.S.C., he moved to the University of Oregon where he has been ever since. He is the author of more than 200 research articles and books and works in Differential Geometry, Global Analysis, Algebraic Topology, and Linguistics. He serves on the editorial board of J. Differential Geometry and Applications, Results in Mathematics, International Journal of Geometric Methods in Mathematical Physics, J. Geometric Analysis, and J. Fixed point Theory and its Applications. He is currently the President of the InterInstitutional Faculty Senate of Oregon and has received the UO Westling Award in faculty leadership.

STANA NIKČEVIĆ

Stana Nikčević received her Ph.D. from the University of Belgrade, at the Mathematical Faculty, under the direction of Neda Bokan. She has been working at that University since 1974. During this period she also sporadically worked at the University of Banja Luka (Bosnia and Hercegovina) and at the Mathematical Faculty in Kragujevac (Serbia). Her research mainly focuses on Differential Geometry. She has maintained international cooperation and has gone on short visits to TU Berlin, Charles University Prague, Universitate Pierre et Marie Curie(Paris VI), the University of Oregon (USA), and the University of Santiago De Compostela (Spain).

RAMÓN VÁZQUEZ-LORENZO

Ramón Vázquez-Lorenzo received his Ph.D. from the University of Santiago de Compostela in 1997 under the direction of E. García-Río and R. Castro. His research focuses mainly on Differential Geometry and he has organized several courses and conferences on Riemannian and Lorentzian geometry.

Index